Changing Sc

A Social and Spiritual Vision for the Year 2020 and Beyond

Edited by Robert L. Menz

University Press of America,® Inc.
Lanham · Boulder · New York · Toronto · Plymouth, UK

University Press of America,® Inc.
4501 Forbes Boulevard
Suite 200
Lanham, Maryland 20706
UPA Acquisitions Department (301) 459-3366

Estover Road
Plymouth PL6 7PY
United Kingdom

Printed in the United States of America
British Library Cataloging in Publication Information Available

Library of Congress Control Number: 2009922308
ISBN-13: 978-0-7618-4576-8 (paperback : alk. paper)
ISBN-10: 0-7618-4576-3 (paperback : alk. paper)
eISBN-13: 978-0-7618-4577-5
eISBN-10: 0-7618-4577-1

™ The paper used in this publication meets the minimum requirements of American National Standard for Information Sciences—Permanence of Paper for Printed Library Materials, ANSI Z39.48—1984

Contents

University Press of America,® Inc.
4501 Forbes Boulevard
Suite 200
Lanham, Maryland 20706
UPA Acquisitions Department (301) 459-3366

Estover Road
Plymouth PL6 7PY
United Kingdom

Printed in the United States of America
British Library Cataloging in Publication Information Available

Library of Congress Control Number: 2009922308
ISBN-13: 978-0-7618-4576-8 (paperback : alk. paper)
ISBN-10: 0-7618-4576-3 (paperback : alk. paper)
eISBN-13: 978-0-7618-4577-5
eISBN-10: 0-7618-4577-1

™The paper used in this publication meets the minimum requirements of American National Standard for Information Sciences—Permanence of Paper for Printed Library Materials, ANSI Z39.48—1984

Introduction

Robert L. Menz, DMin

Every society has certain forms of organization, rules, and norms concerning such institutions as marriage and family, education, religion, economics, and politics. These institutions facilitate the socialization process thus shaping individuals and society as a whole. In shaping, or even controlling individuals, these social forces provide us with predictable patterns that help give life meaning, even as they evolve. Our interactions with them are enduring and continual. They are present when we are born and will continue when we die.

Changing Society: A Social and Spiritual Vision for the Year 2020 and Beyond will provide an overview of traditional as well as other important institutions that are having a great impact on/in the United States and beyond. As this book's subtitle suggests, social and spiritual components such as values, essence, meaning, and purpose will be discussed. *Changing Society* is unique in that social institutions are addressed by writers with keen social insights and great spiritual awareness as they discuss social dynamics and future issues. Consequently, *Vision 2020* in the title suggests more than a calendar date.

Authors of the following chapters discuss both what is traditional and what is new for their respective topics. They present matters that are exciting or alarming, helping or hurting, and then consider the social and spiritual impact of each. Perhaps the most significant part of each chapter is its writer's predictions of where each of these social institutions will be in the year 2020 and beyond. Experts on their topics of each chapter, authors were invited to maintain an attitude of "What I wish others knew about what I know." The reader will find that each chapter is built on solid research and not just personal musings.

Considering the scope of the topics for the following chapters, and the diversity of the authors, the range of issues is necessarily limited and selective. In spite of that fact, the reader will find the content provocative and inspiring.

In the first chapter, Dr. Robert Menz presents the family as the most influential and fundamental institution in society. The family is the first classroom, the first sanctuary, and the first setting where political dynamics are experienced and economic principles—both good and bad—are learned. Of five traditional social institutions, the family is preeminent in shaping who we are. A parent is, in fact, a systems manager responsible for that complex system known as a family. In addition, parents are shaping human beings. A contractor may construct a solid and beautiful house, consistent with blueprints, or, alternatively, construct a building seemingly void of planning with compromised material and structural flaws. A parent may consider a family vision complete with plans and directions for each child's equipping in a complex world where love is known and boundaries are honored, or a parent may proceed with a hodgepodge of reactions and wonder why there is chaos. The more settled the parents, the more settled the child. The more secure and harmonious the child, the more poised he or she is for spiritual, emotional, mental, and even physical growth. To help equip parents, Menz offers insights including articulating inner values and priorities, giving the gift of love, and exemplifying strength and faith. Indeed, children do what parents do. Without a vision and intentionality, one will proceed on autopilot and do what one's own parents did and not what the current situations calls for.

One must acknowledge that the family is undergoing change. It is not the same as it was twenty years ago and certainly not where it will be in 2020. Yet one can be assured the family is not going away. It will continue to be the social institution, the primary group, and the fundamental force that shapes us.

In the second chapter, Dr. James Gebhart conducts us through an enlightening tour of Delaware, Ohio, a prototypical U.S.A. town, representative of mainline religions familiar to most of our readers. Most of us had some sort of understanding of religion when we were growing up. There was comforting predictability because change was slow. To be sure, a Roman Catholic and a Protestant may have had a theological wrestling match from time to time, but for the most part, they could use the same terminology. Today, we may not only learn of Jesus, we may hear of Buddha, Mohammed, or even some expressions of faith without religious figures. The subject of religion therefore can fill large books, even exhaustive volumes. Gebhart masterfully avoids diluting the topic by limiting his focus to rapid changes and polarization within Christianity.

Change in the Christian community is emerging quickly and in some ways with fierce energy. Gebhart proceeds with an awareness of anthropology, history, science, theology, and politics. Then, using his insights as a clinical psychologist, he interweaves those changing dynamics for the reader. He concludes with some forecasts about the continuing romance between religion and politics.

Chapter 3, Education, addresses the concern that students in the United States are falling behind those of other countries in math, science and engineering. Ms. Mazurowski, an attorney and former public school teacher, details the recent history of education in the United States, noting that politics and political beliefs have played a large role in our educational process. She further details current issues in education including: the No Child Left Behind Act; use of tax vouchers to allow children of lesser means to attend private schools; growth of the charter school movement; use of technology in the public schools; privatization of some public schools; home schooling; and economics and poverty in the public schools. The writer also emphasizes the issue of unequal funding among public schools.

The reader will appreciate Mazurowski's vast knowledge of both education and law as she sketches the results of numerous lawsuits brought in various states to ensure equal funding to all schools in those states. Current debates and legal issues regarding recitation of the Pledge of Allegiance and the teaching of evolution in the public schools are also addressed. The chapter concludes with projections of what changes in education we may see by 2020.

The chapter on social and spiritual aspects of Politics sets forth two megatrends: the rise of spirituality and the movement of the center of American politics toward the far right. Within those trends one can detect several lesser, yet quite profound, trends that have led to our present state.

Based on his varied experience as a sociologist and theologian, Dr. Roy Godwin shows how historical developments have shaped the diverse ideologies at play within the realm of American politics. The author's projections suggest a movement back toward a center that will be stationed still toward the right of what used to be, yet away from extremes, with the evangelical and fundamentalist influence waning as the center becomes more dynamic in its national leadership.

Moral issues are here to stay in the national debate, for which the center is mostly grateful, so long as personal morality is not enforced upon the citizens of our democratic republic. Godwin offers that solutions to the gravest issues in our society will require all sides to work together toward solutions by practicing civil dialogue and bipartisanship.

In Chapter Five, Ms. Michael and Mr. Strayer, professors at Edison Community College in Piqua, Ohio, seek to give readers a clear understanding of basic economics in the United States. They review basic principles of economics, including the interaction between scarcity and choice. They trace the historical roots of classical economic theory, starting with Scottish economist Adam Smith who saw individual decision making, self-interest, and the marketplace interaction of supply and demand as foundations of capitalism. Strayer and Michael examine how the role of government in the U.S. economy has evolved from minimal participation to a very active role—one in which government spending has become its second largest sector. The role of government and the size of its participation in the economy are controversial, with critics debating our rulers' ability to deal with economic problems that impact families; they ask if the role of government has become so large as to contribute to the financial problems many individuals and families now face. Their discussion, "The Soul of Economics" guides the reader from a quantitative view of economics, to a qualitative one.

Part Two of the book is opened by Dr. Robert Menz with an informative discussion of Health Care, a topic that has emerged as a major social concern. From a vantage point outside the medical enterprise one may conclude, "It is the best of times and the worst of times," characterized by contrasting realities. First, people in the "West" are living longer and many of the people on our planet are living better. The other reality is that great numbers of persons in our world are not living well at all, and indeed are not healthy.

Menz points out that worldwide wellness cannot be achieved mainly by medical treatment. Good nutrition, safe drinking water, window screens, vehicle seatbelts and helmets, and workplace safety and ergonomic practices will do more for health than the pills and surgeries at our disposal. Events in the political and economic arenas also impact wellness. Positively, there is a correlation between higher levels of education and increased levels of wellness. Negatively, the presence of bigotry, hate, and wars in the name of God, may be harmful to one's health.

With "globalization" and our increased exposure to ideas from all societies, an awareness of "best practices" is emerging. The medical establishment will progressively become less threatened by practices outside their paradigm. Western medicine has followed the pattern of ridiculing supplements and neglecting integrative concepts. With increasing influence of "Eastern" practices, the entire world is poised to benefit from the holistic concept: wellness of mind, body, and spirit.

In "Race and Ethics" attorney Brian C. Thomas reveals an enlightening panorama of racial and ethnic dynamics in the United States. Thomas writes out of his extensive legal background to capture the complex issues of race and ethnicity. Race is among the first things that one may notice about another person, but that is not itself the problem. When people attribute certain negative characteristics to a racial or ethnic group and then consequently mistreat its members, problems often arise.

In this chapter, Mr. Thomas looks at discrimination in the United States—past, present, and future. Racial and ethnic intolerance, hatred, prejudice, discrimination and inequality offend the collective sensibilities of a healthy society. Pointing to the humanitarian vision of Dr. Martin Luther King, Thomas offers us a glimpse of what life would be were we to transcend the racial and ethnic animosities that keep our society from realizing its full potential.

Readers will intuitively agree that changing technology will impact the topics of each of these chapters. Indeed, technology does not exist in a vacuum; its changes pervade our culture, influencing each individual and all of society. In Chapter 8, Dr. Gwen Ogle takes you on an incredible voyage as she explores the role technology will play in the year 2020. But visions of flying cars, holograms, and X-ray vision seem to be science fiction rather than an imminent reality. What can we realistically expect? Technology has affected every facet of our everyday lives, from ways that are so embedded as to be unnoticed, to visible practices that are so pervasive they create a cultural paradigm shift.

As Dr. Ogle explores technology's influence on our social, spiritual, and global world, she points out how it will impact the issues set forth in all the preceding chapters. She assesses its potential impact on each of these areas in 2020 . The irony is that the predictions made by Ogle focus less on what technologies of the future will, can, or should do, but instead on how humans need to understand themselves to best appropriate the benefits that technology can offer.

It has been said that "Hindsight is 20/20." Of course, no one has the ability to speak with absolute certainty about the future. These authors, however, have used historical trends and solid literary research to develop their topics and project where they believe our Social Institutions will be in 2020 and beyond. They invite the reader to enjoy their efforts now, and perhaps again—in the year 2020.

Part I

The Social Institutions

> Western society for the past 300 years has been caught up in a fire storm of change. This storm, far from abating, now appears to be gathering force. Change sweeps through the highly industrialized countries with waves of ever accelerating speed and unprecedented impact. It spawns in its wake all sorts of curious social flora. . . . A strange new society is apparently erupting in our midst. Is there a way to understand it, to shape its development?
>
> (Toffler 1970, 9-10)

Chapter 1

Marriage and Family

ROBERT L. MENZ, DMIN

> In all cultures, the family imprints its members with selfhood. Human experience of identity has two elements: a sense of belonging and a sense of being separate. The laboratory in which these ingredients are mixed and dispensed is the family, the matrix of identity.
>
> (Minuchin 1974, 47)

What is a Family?—A Changing Paradigm

Ask people to define family and you will get varying responses. Beyond the fact that many would describe their current family, or the one that they grew up in as the prototype to define the family, most would agree that the old stereotypical model ceases to capture the modern complex and pluralistic system the family has become.

The traditional and *nuclear* family will here be defined as a social unit consisting of one or two parents rearing their children. The *extended* family is a group of persons of common ancestry. The progressively common *blended* family is a group resulting from joining previously established family units and their children. A family is a social group, a social system, and a social institution. Indeed, the family provides the fundamental setting for the formation of one's meaning and worldview. Most influential of the five social institutions (the other four are discussed in chapters 2-5), the family is a primary network for relations. It may be ideally intimate (close and nuclear), usually intergenerational (apart and extended), and progressively diverse (step and blended).

As a group, the family is changing from a *Father Knows Best* model of the 1950s to accommodate a sufficiently functional *Brady Bunch* model at the close of the century. In today's family, either mom or dad may or may not be present. Either or both may be married (not necessarily to each other). Either or both may have been unmarried (divorced), remarried, or never married. Parents may now create or assemble families that may be interracial, cross cultural, and age-extended. The mother may have a more demanding job outside the home than the father has. In fact, a family now may have a mom or dad who has a partner of the same sex. With these dramatic changes in the last few decades, one may conclude that the family is collapsing. But not so fast with the extinction thinking. It would seem that change isn't destroying the family as much as the family seems to be surviving in spite of the changes.

The family is also a system. If we consider a family that consists of mom, dad, and two children, we must also consider the entity beyond each of them. The conglomerate body (the family) is shaped by each one of the family members with each individual strictly controlled by his or her family role. The family as a system influences all of its members but is constantly being influenced by other and greater systems as well. Just as the human body has many systems, such as the nervous, immune, cardiovascular, digestive, and skeletal, the family is comprised of many subsystems which contribute to the whole. In addition, we are all influenced by educational, religious, political, economic, and other systems as well. As it becomes progressively difficult to avoid the impact of global forces, the family system is certainly impacted by them. Along with the changes that the family has encountered over the past several years is the expanding perspective or paradigm shift of the average U.S. resident. The traditional white Anglo-Saxon Protestant (WASP) view of marriage and family was of an intact nuclear family, extending back over generations. As we have come to appreciate, the WASP perspective was once dominant in influence if not in numbers; it certainly is not dominant now. As Irene and Herbert Goldenberg say in their book *Family Therapy*:

> [African Americans] expand their definition [of family] to include a wide informal network of kin and community. Italians think in terms of tightly knit three-or-four-generational families, often including godfathers and old friends; all may be involved in family decision making,

> may live in close proximity to one another, and may share life cycle transitions together. The Chinese tend to go even further, including all their ancestors and all their descendants in their definition of family. . . . Native American family systems are extended networks, including several households. A non-kin can become a family member through being a namesake of a child, and consequently assumes family obligations and responsibilities for child rearing and role modeling. Hispanic Americans, the fastest growing ethnic group in the country, take deep pride in family membership, with a man generally using both his father's and mother's name together with his given name. (1996, 36-37)

Indeed the family is undergoing changes, yet our awareness of diversity is expanding even more. Perhaps the *Ozzie and Harriet* model existed more in fantasy than in reality. Perhaps the *Ozzy Osbourne* family model has been among us more than we've openly admitted. One seeking the typical traditional family of fifty years ago, will find the model particularly illusive in today's United States.

Evolving Statistics

A Family Profile

According to the U.S. Census Bureau, "[A] family is a group of two people or more (one of whom is the householder, [that is to say, the head of the household]) related by birth, marriage, or adoption and residing together; all such people (including related subfamily members) are considered as members of one family" (U.S. Census Bureau 2004). The Bureau also defines households. "A household consists of all the people who occupy a housing unit" (2004). So a household may or may not be family. In fact households may be "family" or "non-family." As one might expect, the last part of the twentieth century revealed a dramatic increase of households consisting of single-parent families and persons living alone. A report prepared for the Census Bureau by Ken Bryson showed that the greatest decline in the proportion of households composed of married couples and the greatest increase in one-person households occurred between 1970 and 1980 (Bryson 1996).

Of all households today, seventy percent are family households and thirty percent non-family. In 1950, eighty-nine percent of all households were family and only eleven percent non-family. Now the average family size for married couples is 3.25. It is 2.85 for a family with a male householder and 3.02 for a family with a female householder. The average size of a household (family and non-family) is smaller at 2.64. Compare this to the average size of a household of 5.4 in 1790, 4.2 in 1900, 3.3 in 1940, and 2.64 in 1997 (Eshleman 2000).

In 1970, married couples with children made up forty percent of households; in 1995 they made up twenty-five. In 1970, people living alone made up one-sixth of all households; in 1995, one-fourth. In 1970, forty-four percent of families had no children under age eighteen at home; in 1995 the figure was fifty-one percent. Also during this same time frame, mother-led families (with no husband present) jumped from 5.6 to 12.2 million and father-led families (with no wife present) jumped from 1.2 to 3.2 million (Bryson 1996).

The nuclear family of the twentieth century seems to have become the prototype by which families are measured. The traditional family became synonymous with this nuclear model. If one broadens the view of family over a greater historical time frame, that nuclear family may be more the exception than the rule. Even the eighteenth and nineteenth centuries saw households that were blended and extended. One house may have had Uncle Joe and his three children within because Joe's wife died in childbirth. Another may have had Susan and her four children because her husband died in the war—and so it went. Ideal models are ideal; yet, they are still models.

The face of the traditional family is changing. Since 1960, the marriage rate has sharply declined and the divorce rate has more than doubled. In 1974, divorce replaced death as the principle cause of family dissolution. Between 1960 and 1998, the percentage of single-parent families more than tripled. Cohabitation without marriage increased from under a half-million in 1960 to over five million in 2000 (Bennett 2001).

At the beginning of the twenty-first century, quite unlike the situation one hundred years earlier, almost all families totally dependent upon specialty domains to help them survive. The typical family today does not grow or prepare its own food, raise and butcher its own meat, nurse and heal its own wounds, or repair its broken household items or its vehicles.

Today's relationship between the family and certain social institutions—particularly religion—is evident. Many families pride themselves on being a Jewish family, or having a Muslim home, or being good Christian parents. Historically, and even today, some families have assumed their religious perspective to be the best, if not the only one. Another example of how social institutions impact each other is the changing economy of the post-World War II period which drew women, including married women, into the workplace. Fifty years ago social norms exerted pressure on wives and mothers to work at home. Today the norm is for mom and dad both to be in the workforce. Granted, this expectation is still stronger for men, yet women are beginning to feel the social pressure to be productive outside the home as well.

During the last half of the twentieth century, women gained greater control over their destiny. Contraceptives like "the Pill" were more effective, abortion was legal, daycare was available, and employers were hiring. The traditional home in the suburbs, with a garage and a pet, now for the first time may sit empty most of the day. Whether women are working outside the home to pursue personal goals or to provide greater family income, the birthrate is affected. Robert Reich reports, "Even women who plan to have children are delaying. Births to teenagers have dropped dramatically, reaching by the year 2000 the lowest rate in the United States since the government began tracking births in 1906" (2000, 165).

There is a decline in the traditional view of the family for a pair of reasons. One is that the rate of marriage has been decreasing and the other is that the rate of divorce has been increasing. In 2004 there were 2.160 million marriages. Numerically, the figures have been roughly stable since 1975. Dividing this number by the total U.S. population, the rate is 8.5 marriages per 1000, a number consistent with 1960 levels. However, when the rate is calculated using the number of never-married women age fifteen and older (those eligible for marriage), it is evident that the marriage rate began to plummet around 1975. Figured in this more accurate way, the marriage rate has fallen from about sixty-five per 1000 never-married women age fifteen and older in 1970, to an all-time low of about forty-six per 1000 today. Viewed this way, the marriage rate has fallen more than forty percent since 1970 (Tishler 2007).

Today's family members are more educated, have more diverse interests, and live longer than the stereotypical, traditional family. Families that include mom, dad, and three children have yielded to single-parent families, blended families, commuter families and "empty nest" couples who have the potential of being together longer without their children than they were together with their children. This being said, the family remains. The changes of the past fifty years have been great. The way it used to be, or perhaps the way we wish (or thought) it was, is no more. Even as it changed, the institution has demonstrated its strength and endurance through the millennia. The family will continue to survive.

A Marriage Profile

For the first half of the twentieth century, the trend in the United States was toward an earlier age at first marriage and a narrowing of the age difference between bride and groom. The second half that century, however, revealed a gradual increase in the median age at first marriage (Eshleman 2000).

The traditional U.S. view of marriage historically included several forces that are less powerful today. The endogamous nature of the traditional marriage—marrying within one's own group—as well as external support of law, religion, and social custom, contributed to the permanence of the union. Marriage may now be legally dissolved without reason or fault (such as desertion or adultery); in fact, some are opting for their union to occur without legalities. Today unmarried couples can live together, unmarried mothers can and do keep their children, and abortion is legal. In a country where autonomous individuality is pursued and celebrated, the man's position as dominant head of the family and its primary provider no longer goes unchallenged. Today it is very common for a woman to be in the workplace, and delay both marriage and childbearing. In fact, it is socially permissible to be single or childless. Traditionally, marriage was carefully defined and regulated by society. Each partner had specific roles to play. One or both parties in the marriage were likely shunned by society if social norms were violated. Today the social attitude seems to be "to each his own."

The practice of unmarried couples' living together out of wedlock has increased dramatically in the past twenty years. In 1988 about nineteen percent of married couples reported that they lived together before they married. Most recently, it has nearly doubled to thirty-seven percent. The number of unmarried households grew almost eightfold from 523 thousand in 1970 to 5.8 million in 2004. The twenty-to-twenty-four age group has the highest percentage of unmarried cohabitation (Tishler 2007).

Religion has greatly influenced the institution of marriage as well. The New Testament verse, "Wives, submit yourselves unto your own husbands, as unto the Lord. For the husband is the head of the wife, even as Christ is the head of the church" (Ephesians 5: 22-23), is not upheld today in mainline Christian churches, as it is in more fundamental Christian denominations. Indeed, Hebrews, Christians, and Muslims all draw from the same source which says that Adam was first (in every way). Eve was created to provide companionship and help to Adam. Norms—and in some cases laws—supporting such views were solidly established. For good or for ill, the effects of these strong teachings are still evident.

Even though it is a universal social norm for marriage to be between one man and one woman, exceptions occur. Since the 1970s gays and lesbians have been progressively revealing their relationships. Despite the fact that marriage is the social legitimization for sex and offspring, couples are cohabitating without a marriage license and having children out of wedlock. Studies show that unmarried cohabitation arrangements are, overall, less stable than marriages. The likelihood of a first marriage's ending in separation or divorce within five years is twenty percent—thirty-three percent within ten years. In contrast, the probability of a premarital cohabitation's breaking up within five years is forty-nine percent—sixty-two percent within ten (National Center for Health Statistics 2002). Although marriage is still intended to be "until death do us part," approximately fifty percent of marriages end in divorce. "Starter marriages" is a new term that has recently appeared on the matrimonial horizon. The term plays on the expression "starter house"—the place one moves into for an intended short time. In her book, *The Starter Marriage*, Pamela Paul discusses these marriages that last for five years or less and end without children (2002). Even though this concept seems to fly in the face of most couples' intentions, if a divorce is inevitable it is usually less complicated when it occurs after a brief marriage that produced no children.

A 2006 news release from the National Center for Health Statistics reported that a third of men marry by age twenty-five and two-thirds marry by age thirty. Half of all women are married by age twenty-five, and three-quarters by age thirty. On the whole, men marry later than women and a woman marries a man two years her senior. Of men who marry as teenagers, fifty percent are divorced within ten years. Yet for those men who waited until they were twenty-six or older, only seventeen percent were divorced within ten years (2006).

Divorce Statistics

When we hear that the divorce rate in the United States is about fifty percent, it is reasonably accurate data but it fails to capture the full picture. For any given marriage to have a fifty-fifty chance depends on many variables. This statement assumes that current trends will continue. Even more pertinent are the details of distribution. According to *Divorce Statistics*, a web site that tracks various divorce data (2006a), the divorce rate for first marriage is forty-one percent. The divorce rate for second marriage is sixty percent and that for third marriage is seventy-three. Yet there is even more to consider. The rate for couples with children is forty percent and the rate for couples without children is sixty-six percent. The age at marriage also is a determinant for divorce. Those marrying under age twenty have the highest rate of divorce. For people marrying between the ages of twenty and twenty-four, there is a 36.6 percent rate of divorce for women and 38.8 percent for men. After that, each five-year increase in the age at marriage yields a decreasing divorce rate until marriages between thirty-five and thirty-nine produce a rate of 5.1 percent for women and 6.5 for men. This general trend was already prevalent in the 1950s (Landis 1958). Geographically, Nevada has the highest divorce rate: 9.1 per 1000 population, and Massachusetts the lowest, at 2.4.

Looking closely at divorce reveals even more variables. According to research done by Barna Research Group in Ventura, California (1999), Baptists have the highest divorce rate of any Christian denomination. The article reported that twenty-nine percent of all Baptists have been through a divorce. Catholics and Lutherans have the lowest divorce rate: twenty-one percent. Interestingly, atheists and agnostics have the same relatively low rate. The report also stated that whites are more likely to divorce (twenty-seven percent) than African-Americans (twenty-two percent), followed by Hispanics (twenty percent), and Asians (eight percent).

How do the divorce statistics in the United States compare with other countries? According to *Divorce Magazine* (2006), when divorces as a percentage of marriages are considered, Belarus, Russian Federation, Sweden, Latvia, Ukraine, Czech Republic, Belgium, Finland, Lithuania, United Kingdom, and Moldova all have higher divorce occurrences than the United States. However, when the divorce rate per year for each 1000 population is considered, the United States has the highest divorce rate: 4.95. Russia has 3.36, United Kingdom has 3.08, Sweden, 2.4,

Japan, 2.2, France, 1.9, Finland, 1.85, and Italy 0.7. Canada is very similar to the United States with an over-all divorce rate of about 48 percent (Divorce Statistics 2006b). Divorce rates in India are among the lowest in the world: 1.1 percent (Divorce Statistics 2006c).

Why are Marriages Failing?

To attempt to understand the contrast between the divorce rate in the United States and that in India, one must look beyond the issues of marital dynamics. There are likely powerful social and religious forces at work here. In the United States, we embrace a "class system" where one is given social permission to pursue greater things and move up the socio-economic ladder. India, on the other hand, has barely moved beyond the Caste System that existed for centuries before the 1950s, and in which one's place in society was immutable and the possibility of changing social status impossible. This is further reinforced by the religious notion of reincarnation. That view holds that to the degree people are faithful to the norms of society and the state of their own condition, they are properly aligned with their spiritual purpose and can expect to improve their condition in a later life. In the United States, if a group does not like what is happening in its place of worship, it is free to leave and establish a new place of worship.

In a similar way, the Japanese have far fewer occurrences of divorce than the United States. Workers there demonstrate loyalty to their employers now virtually unknown in the United States. Japanese workers may not like their jobs any more than U.S. workers, yet they will likely stay in their jobs whereas U.S. workers will not. Consequently, and to the point, the U.S. culture more readily allows for relationships to be severed at work and—not incidentally—at home.

One obvious reason that the marriage survival rate has decreased over the past fifty years is the expanded social exposure granted to individuals by our society (and other societies of the West). The options available to an individual (or to a couple or to a family) are profoundly different from the situation half a century ago. Such expanded opportunities often create a sense of restlessness, even a temptation to test the new—perhaps to climb the next mountain simply because it is there; to go where one has never gone before, so to speak.

I once heard the story of a five-year-old named Sally who asked her parents where she came from. Mom and Dad thought this inquiry a bit premature yet were prepared and consequently went into detail about relationship, love, pregnancy, and the birthing process. Sally responded by saying, "Well, my friend Allison comes from St. Louis."

It takes many of us a while to say where we came from. Our mobile society is causing families to lose a sense of "where they are from"—where their roots were. Seventeen percent of Americans change residence annually. Forty percent of second graders have already attended another school. About three percent of families move to another state each year (Reich 2000), and as we have already seen, many are changing spouses and family members.

Changes in family structure seem to be reflected in society at large. There was a time in the mid-twentieth century when large manufacturing companies offered their employees predictability and security—the assurance of employment. Such assurance and dependability seem to have been replaced with the expectation that workers must "earn their wings every day." Continual—and sometimes sudden—change characterizes modern life, requiring us not only to adapt but to prepare for even further changes.

Another factor that contributes to marriages' not surviving is social permission. The stigma associated with separating, divorcing, or cohabiting has been greatly minimized. The social "ice" was sufficiently broken more than thirty years ago, so that today a "failed" marriage is no longer headline news.

In the last half of the twentieth century, the United States and many other parts of the world saw women seeking and winning profound changes. Women gained a greater voice at work and at home. Women today enjoy more political power and financial freedom than ever before. Men, however, have not experienced such radical advances in recent decades. This disparity has the potential to intensify conflict within a marriage. In *The New Rules of Marriage*, Terrence Real suggests that these societal changes call for a new marriage. In this new relationship, partners' actions can generate greater intimacy but, at the same time, require greater sacrifice than the twentieth-century model. Real suggests that marriage partners must each ask the question: "What can I give you to help you give me what I want?" This practice Real calls "The Golden Rule of Relationship Empowerment (2008, 7). When marriage partners engage in this practice, their relationships can indeed become stronger.

Positive change needed in a relationship will come when both partners become intent on fairness and improvement. They come to realize that each of them is responsible for bringing control and change to their shared relationship. In her book, *The Proper Care and Feeding of Marriage*, Dr. Laura Schlessinger says that one of the most important lessons a couple can learn now is: ". . . the incredible power of making the other feel cared about, special, important, valued, admired, loved, and appreciated as a real woman or a real man" (2008, 106).

Another dynamic, particularly in the Western hemisphere, involves the issues that result from romantic choice of a marital partner. Although this option is a relatively new one, many people, especially in the United States, presume that our system of selecting a mate is the way it always was and the way it is elsewhere. But this is incorrect. Not only is the romantic system of mate selection relatively new, but in much of the world, arranged marriage is still the acceptable way for couples to unite.

For the past fifty years, the "systems" approach to marriage and family theory and practice has progressively shed light on the subtle dynamics of such things as attraction, complementary balance, and responsive behavior. This approach to the understanding of marriage and family dynamics has been informed and shaped by such eminent scholars as Murray Bowen, Jay Haley, Salvador Minuchin, Virginia Satir, and Carl Whitaker. One family systems therapist relevant to this discussion is Harville Hendrix (1988). Hendrix proposes that a person becomes attracted to one whom he or she unconsciously perceives as having the potential to make him or her whole. This freedom of selection is scarcely an option, of course, with arranged marriages. But people who have the option of selecting their own mate will unconsciously seek one who will help heal childhood wounds and make them complete and well again. Hendrix suggests that we gravitate to one who has the potential of developing a family system similar to the one of our origin—the one where our wounds occurred. Consequently we marry a person who is quite capable of re-opening our wounds. Certainly this is not our conscious plan, nor is it consistent with the erotic feelings that accompany the unconscious promise of need-fulfillment.

Paradoxically, what brings couples together often drives them apart. For instance, the woman who said she was attracted to her husband because he was so "laid back," later wanted to divorce him because he was unmotivated. There was also the man who was attracted to his wife because her hair looked so beautiful and later became upset because she spent so much time arranging and caring for it. Then there was the one who found the future spouse free and spontaneous and soon threatened that there would be an ex-spouse if the partner did not embrace more organization and structure. Certainly, in today's world, this option is often chosen. As one's partner reminds one of the family of origin, or perhaps of the beloved person "who married dear ole' Dad (or Mom)," couples often later find themselves saying things like, "Don't mother me," or "You sound just like my dad."

As we moved from arranged marriages, to romantic, unconscious marriage choices, the divorce rate has increased to where it is today. Because of this, Harville Hendrix has proposed the need for a third approach to marital arrangement—the Conscious Marriage. In his book, *Getting the Love you Want: A Guide for Couples* (1988), Hendrix offers a systematic, therapeutic response to the problems and conflicts of many contemporary marriages. It is a very involved therapeutic process, but simply put, it helps couples learn how to give each other what is needed in order to facilitate the healing process of the wounded child. The success rate of what Hendrix calls *Imago Therapy* has led many marriage counselors to learn this method.

Another negative force exerted on families now is a financial one. Elizabeth Warren and Amelia Warren Tyagi in the book *The Two-Income Trap* (2003), reported earlier in this decade that one of the most significant burdens on family life is financial stress. For example, couples with children are twice as likely to file bankruptcy as their childless counterparts. To emphasize the significant impact the financial pressures are having on some families, the authors state that in a year's time:

> More people will end up bankrupt than will suffer a heart attack. More adults will file for bankruptcy than will be diagnosed with cancer. More people will file for bankruptcy than will graduate from college. And, in an era when traditionalists decry the demise of the institution of marriage, Americans will file more petitions for bankruptcy than for divorce. Heart attacks. Cancer. College graduations. Divorce. These are markers in the lives of nearly every American family. And yet, we will soon have more friends and coworkers who have gone through bankruptcy than any one of these other life events. (2003, 6)

Warren and Tyagi report an alarming six hundred percent increase in bankruptcies over the past twenty years. Furthermore, it is reasonable to conclude that for every family that files for bankruptcy, there are many other families undergoing major financial difficulties. It is also alarming to consider that such a great portion of the bankruptcies are being declared by middle-class families in the United States. The authors declare, "By every measure except their balance sheets, the families in our study are as solidly middle-class as any in the country. And they are united by another common thread: Most of these families sent two parents into the workforce" (2003, 7).

So are we talking about a growing trend of irresponsibility? Are many families making bad financial choices? Is it simply a matter of overspending? In some cases, yes. Yet there are families whose "bread winner" was called to active duty, or was a victim of crime or natural disaster. Nine out of ten families with children cite these reasons for bankruptcies: job loss, family breakup, and medical problems (Warren and Tyagi, 2003).

Future Trends—2020

One must acknowledge that the family has changed and will continue to change. This awareness does not imply only negative change nor does it imply that the family will become extinct. Yet, certainly, the family is much more diverse and complex than it used to be.

Traditionally, the marriage was to be endogamous, legally filed, between one man and one woman, with premarital chastity and sexual exclusivity, husband-lead, wife-supported, and separated only by death. In 2020, there will be much more of the non-traditional family models we see today including the choice to not marry or to not have children. There will progressively be multiple marriages, same-sex unions, more interracial and interclass marriages, more dual careers, commuter marriages, and certainly step and blended families. In fact, in 2020 there will be those who seek out biological siblings who, like themselves, were conceived through donor sperm. It is conceivable for one to learn that he or she has 20 (or more) "half brothers" or "half sisters" due to the common sperm source that fertilized one's mother's egg. These people would be genetically related, yet the more common step or blended family model would not capture their (family) relationship. These changes will certainly be stressful and reveal breaking points. Yet this diversity will provide a cutting edge to broaden perspectives, enhance tolerance and expand personal durability.

In the year 2020, the family will continue to be the primary social institution for teaching and assimilating societal values. Having an intimate, primary group—the family—(even with a broader face) will continue to be important. The family will continue to offer a sense of safety, belonging, and love that one cannot find in other parts of society. The institution called "family" will continue being diverse and perhaps harder to define. Yet the adaptive nature of the human spirit will respond with greater levels of resilience and perseverance.

For reasons in addition to personal, romantic choice, divorces are increasing and families are separated more frequently. Because the institution of family is woven into the fabric of other social institutions, a change in any will reverberate through all. Some have personally reported that their failed marriage was a tragedy. Some have responded by saying their divorce was the solution—the best course of action for all involved. From a given perspective, an ended relationship can seem good or bad, right or wrong. One thing is certain: we are all in these rapidly changing systems together. Because of our shared need to survive, our society will progressively become acclimated to these changes. In 2020, agencies helping families will still exist. However, there will be no government service, no social service agency, no charity group, no security system, no financial conglomerate, and no nursing efforts that will come close to doing what the family has done and will do to help itself. In traditional and blended family models as well as single family homes, members will continue to retreat to their family structures to find acceptance, guidance, discipline, love, and meaning.

Toward Healthy Families

For a family to thrive in a world of rapidly changing norms, parents must engage in several healthy practices. Foremost among these practices is that parents keep the family (and the marriage) at the top of their priority list. This does not mean that there will not be days where such things as job, illness, and other major concerns will consume a person—yet the individual who bounces back is one who considers what is ultimately important. At the end of life, one will not likely regret having made too few widgets (or whatever was done through the career). One may keenly regret, however, not having made the family a greater priority.

A second practice for parents to consider in promoting family health is to foster resiliency. Each generation has its stressors. For this writer's generation it was crawling under the school desk in preparation of a nuclear attack. If the desk did not offer the necessary protection, it was "curtains" anyway. Today, however, the post 9/11 world seems even more precarious. Today it may seem less likely that an entire city would be destroyed than that a whole building or a bus, or a plane might be. Parents who provide a forum to talk about such fears and concerns provide not only security but also processes and solutions to life's difficulties. For everyone in the family to be able to bounce back and recover from a stressful event is vital in life. Today, and increasingly into the future, healthy families will need to have the emotional resources to navigate through very stressful events.

A third healthy practice for parents to embrace is to realize that they teach by what they do. If what a parent says is consistent with what the parent does, there is authority. If the behavior and the words are inconsistent, authority is undermined. For a parent to articulate the need to be honest, but practices dishonesty, or to say "I don't want you to pick up my bad habit of smoking," is to engage in a failed attempt at teaching. Parents and children (regardless of age) are transparent to each other—albeit perhaps in different ways. It behooves us as parents to ask, "Is this behavior what I want to teach my child?" Furthermore, we must teach a healthy balance between sacrifice and gratification, between giving and requiring, between freedom and boundaries, and between grace and discipline.

A fourth practice for parents to develop is marital and family intimacy. Family intimacy goes beyond the horizontal bonds of a relationship to include the vertical and horizontal plane. When one feels connected to others, related to nature, and awed by the sacred rhythm of the universe, one is more equipped to nurture the relationships within the family. This vertical dimension of intimacy recognizes differences and encourages diversity. In fact, respect for uniqueness and autonomy is not only a sign of emotional and relational maturity, but such healthy respect for the other also strengthens relationships within the family. To encourage every person in the family to grow and enhance his or her uniqueness and gifts is a true spiritual exercise that leads to mature closeness. Families and individual members must reach out to the concerns and needs around them. To cultivate friendships, work to meet community needs, and address environmental concerns is to invest in a world beyond oneself. This communal outreach is not only a measurement of family health; it is an exercise to enhance family intimacy.

Another practice to consider in the pursuit of healthy families is to utilize community and spiritual resources. A family in the twenty-first century needs all the neighborly, community, organizational, and governmental "propping up" that it can get. Schools and religious communities offer many helpful activities and supportive relationships. When a child has his or her schedule filled with purposeful extracurricular activities, it leaves little time for un-redeeming ones. Parents need not only community resources to assist with parenting, they also require divine assistance. For parents to acknowledge the presence of and seek assistance from "a higher being" is not only a healthy thing to practice (and thereby teach), it is often a necessary thing to do. Prayer and meditation not only seek divine help and direction, they provide a forum for a quiet mind to process responses and solutions.

A sixth exercise for parents to pursue is to be intentional in feeding the family's intellect, emotions, and spirit as well as feeding each member's physical body. Herein, one again leads by example. Engage in stimulating and intellectual dialogue. Establish and maintain family meal time. Read to younger children—do creative and challenging games and puzzles with the older ones. Seek to understand and label emotions. Teach that emotions are not bad, but that ignoring and misusing them can be. Seek out and comment on ultimate and meaningful experiences. Teach what is revered about our environment—how a sunset may be transcendent—how life, even when painful, is sacred.

Finally, to enhance the family's health, develop and implement a family mission statement and project a family vision. This is not to suggest that a family must write and frame a statement as businesses do. However, there is great merit in discussing your purpose and declaring your intent and goals as a family. A family mission statement will clarify who you are and what your family is about. Knowing who you are, what you're about, and where you are going as a family provides meaning and purpose, while goal setting identifies and highlights accomplishments. How do you want your family to look (and be) in the year 2020?

The Soul of Family—A Summary

In the same way traditional families provided commitment and accountability, identity and support, the families of tomorrow will also. Families may have a different, even a more complex look, but they still must provide these fundamental ingredients for social cohesion.

The family has survived longer than any other social institution. From before recorded history to beyond the horizon of the future, families have existed and will exist. Societies and cultures change, yet, it is the nature of the family to adapt. From ancient nomadic to settled agricultural and back to mobile societies, families have adjusted and will continue to survive.

Families provide the foundation and structure of belonging. To belong to a system that offers caring, accountability, guidance, and acceptance is the gift of the family and the longing of humanity. Family members gain a basic security from belonging to each other and providing a place to love and be loved. This is not to imply that families are free from trouble and pain. Strong families are not those who navigate through difficulties and emerge free from problems. Strong families are the ones that have united during a crisis, weathered the storms, learned vital lessons, and (like broken bones) healed stronger than before.

We face more change now than any other time in history. This trend will only accelerate. Yet the family will continue to be the social institution, the primary group, and the fundamental force that shapes us. Within the diversity of the global community where we find ourselves, the family, more than any other force, will influence the individual characteristics of our lives, the role we play in society, and the world we create for families yet to come.

References

Barna Research Group. 1999. "Christians Are More Likely to Experience Divorce Than Are Non-Christians." www.valleyskeptic.com/christdivorce.html. (accessed August 30, 2006).

Bennett, William J. 2001.*The Broken Hearth*. New York: Doubleday.

Bryson, Ken. 1996. *Current Population Report: Household and Family Characteristics: March 1995*. U.S. Department of Commerce, Economics and Statistics Administration, Census Bureau. October 1996.

Divorce Magazine. 2006. "World Divorce Statistics." www.divorcemag.com/statistics/statsWorld.shtml. (accessed July 18, 2006).

Divorce Statistics. 2006a. "Divorce Statistics in America." www.divorcestatistics.org. (accessed July 18, 2006).

Divorce Statistics. 2006b. "Divorce Rate in Canada." www.divorcerate.org/divorce-rates-in-Canada.html. (accessed July 18, 2006).

Divorce Statistics. 2006c. "Divorce Rate in India." www.divorcerate.org/divorce-rate-in-India.html. (accessed July 18, 2006).

Eshleman, J. Ross. 2000. *The Family: Ninth Edition*. Needham Heights, MA: Allyn and Bacon.

Goldenberg, Irene and Herbert Goldenberg. 1996. *Family Therapy: An Overview, Fourth Edition*. Pacific Grove, CA: Brook/Cole Publishing Company.

Hendrix, Harville. 1988. *Getting the Love You Want: A Guide for Couples*. New York: Harper and Row.

Landis, Judson T., and Mary G. Landis. 1958. *Building a Successful Marriage*. Englewood Cliffs: Prentice-Hall, Inc.

Minuchin, Salvador. 1974. *Families and Family Therapy*. Cambridge: Harvard University Press.

National Center for Health Statistics. 2002. "New Report Sheds Light on Trends and Patterns in Marriage, Divorce, and Cohabitation." U.S. Department of Health and Human Services, Centers for Disease Control and Prevention. www.cdc.gov/nchs/pressroom/02news/div_mar_cohab.htm. (accessed July 18, 2006).

National Center for Health Statistics. 2006. "Fertility, Contraception, and Fatherhood: Data on Men and Women from the National Survey of Family Growth." U.S. Department of Health and Human Services, Centers for Disease Control and Prevention. www.cdc.gov/nchs/pressroom/06facts/fatherhood.htm. (accessed July 17, 2006).

Paul, Pamela. 2002. *The Starter Marriage and the Future of Matrimony*. New York: Villard Books.

Real, Terrence. 2008. *The New Rules of Marriage: What You Need to Know to Make Love Work*. New York: Ballantine Books.

Reich, Robert B. 2001. *The Future of Success*. New York: Alfred A. Knopf.

Schlessinger, Laura. 2008. *The Proper Care and Feeding of Marriage*. New York: HarperCollins.

Tishler, Henry L. 2007. *Introduction to Sociology: 9th Edition*. Belmont, CA: Wadsworth/Thomson Learning.

Toffler, Alvin. 1970. *Future Shock*. New York: Bantam Books.

U.S. Census Bureau 2004. "Current Population Survey (CPS)—Definitions and Explanations." www.census.gov/population/www/cps/cpsdef.html. (last modified January 20, 2004; accessed July 17, 2006).

Warren, Elizabeth and Amelia Warren Tyagi. 2003. *The Two-Income Trap*. New York: Perseus Books Group.

Chapter 2

Religion

James E. Gebhart, PhD

> Nothing finite can cross the frontier from finitude to infinity. But something else is possible: the Eternal can, from its side, cross over the border to the finite. It would not be the Eternal if the finite were its limit. All religions witness to this border crossing, those of which we say that they transmit law and vocation to the peoples. These are the perfecting forces from the Unlimited, the Law-establishing, the founding and leading of all being, which make peace possible.
>
> (Tillich 1966, 63)

It is the very essence of human beings to reflect upon how they stand in relationship to what they consider the divine. The philosopher Huston Smith describes the human as a theomorphic creature, that "every human being has a God-shaped vacuum built into its heart" (Smith 2001). Thus there will always be a religion. The study of its state and future, the task of this chapter, is to extrapolate the trends as we are able to identify them. In many ways this is an impossible task as trends emerge quickly, sometimes with fierce energy, and they change directions just as quickly. These trends are effected by many factors: history, which requires the uses of the past; anthropology, which defines the genuine needs of the culture; science, which challenges and enlightens theology; psychology, which describes the anxiety and interprets the behavior of the people; economics, which addresses the class struggles from exploitation and alienation; and politics. Especially politics, because institutional religion "is always politics by a different name" (Dark 2005), and, likewise, politics easily assumes the features of religion where faith and patriotism are often seen as synonymous. Further, the shifting vicissitudes of politics are made infinitely complex by its global dimensions.

The focus of this chapter must be limited to Christianity in North America as compared to Christianity in Europe and the Southern hemisphere. It will be possible only to give passing reference to two other world religions, Islam and Judaism, as they too are so politically fused. Also Buddhism, Hinduism, and other Eastern groups will not be assessed as they can be expected to continue in their steady state just as they have survived the passing politics of the past.

The Situation in North America

Delaware, Ohio, is a lovely and classic small city of Middle America. It is a county seat where traditional lifestyles are being morphed by urban sprawl from Columbus, the state capital, just twenty miles to the south. Drive into it from Route 23 and the first thing you see on the tree-lined streets which lead to the bustling downtown area are the many fine churches. The Roman Catholic gothic structure is the first on the left, followed by the two United Methodist Churches (remnants of pre-merger Methodist times), the Presbyterian Church, the Lutheran Church, the Episcopal Church, an American Baptist Church, and the United Church of Christ. Here are the representatives of mainline Christianity in a typical Our Town, U.S.A.

These churches are a testament to well over a century of religious commitment. Generation after generation has worked to organize the various faith communities, raise up the strong and beautiful houses of God, and carefully attend to them, year after year. A local church college and a denominational seminary further attest to the character of the community history.

However, significant change is clearly underway. In all likelihood the children of these faithful will not persist in this same devotion and things will not continue as before. A number of polls have been taken to determine church attendance at the turn of this century, and although there is some variance the polls tend to agree: actual church attendance has slipped from forty-five percent in 1966 to twenty- to twenty-five percent today (Pressor and Stinson 1989, 137-145; and Adler 2005, 48-

50). In the United States there is regional variance with attendance being one and one-half to two times higher in the South and Midwest than in the West or Northwest. Further, Pentecostal and some groups defined as Evangelical are reporting growth rates, but they do not have a standardized and reliable form of reporting. Still, if current trends continue, the forecast for the future is for continuous decline in church attendance—down to fourteen percent of Americans by the year 2020 and to less than ten percent by 2050 (Olson 2006, 11-13).

So in Delaware, Ohio, some of the children, now young adults, might continue the tradition of their parents although they might find the music odd and have difficulty with a theology that does not address the post-modern era. Many, however, will not attend church regularly. At the college campus, where chapel was once mandatory, the campus ministries will work to find creative modes for relating to more skeptical students.

Where will this new generation turn in search of a meaningful religion? As is the case across North America, some, put off by "religious simplism" and who question whether Christianity "can survive the inquiries of a lively mind . . . and bear the scrutiny of serious intellectual investigation" (Hall 1993, 5-6) will move to a respectful agnosticism, always open to dialogue. Others will move into what one scholar has called "The Third Disestablishment" wherein organized religion is perceived as no longer providing an integrative force in society, and faith becomes more *individually* important and less *collectively* significant (Hammond 1992, 1-18). These persons will identify with the popular and thriving movement to personally seek communion with the divine in meditation, retreats and gatherings designed to foster transformative experience (Adler 2005). Still others will seek out new assemblies and forms of religious life that offer an alternative, contemporary worship experience, characterized by more informality and modern music. Furthermore, there is the evolution of the megachurches, the large communities with elaborate facilities. The megachurch can consist of several thousand members as it seeks to offer a variety of worship and educational experiences, structuring itself so that everyone can be invited to participate in a small group of persons with similar interests and values. Most but not all of the large assemblies are referred to as Evangelical communities, characterized by conservative religious and social philosophies and fundamentalist theology. Finally, approximately twenty-five percent are seen as moving toward the more radical right, usually characterized as the fundamentalist, evangelical, holiness or Pentecostal, to be discussed below (Noll 2002, 268-283).

This contrast, in Delaware County, Ohio, between the traditional religious structures of the past centuries and the newly emergent assemblies mirrors a phenomenon that is occurring across the United States. At first glance it appears to be the familiar option of diversity which has characterized the development of religious life in the United States. Parents and grandparents of this generation can remember when Roman Catholics and Protestants nervously co-existed, each group seeking to be civil while not really understanding or trusting the intentions of the other. Of course, those Protestant churches at one time regarded one another with considerable suspicion, and a person identified as being a Presbyterian or a Lutheran or a Methodist or a Baptist took pride in those separate and sometimes contrasting traditions and practices. So we are familiar with diversity. But a deeper investigation quickly reveals that the current divisions are not just contrasts but elements of more complex divisions regarding religion on the U.S. scene, religion which has been evolving and driven by history, anthropology, science, and most especially by economics, psychology and politics.

The people of Delaware, Ohio are not likely to be keenly aware that their current religious positions have been shaped by forces from the past which are continuing to evolve. Although many of the founders of our nation were Deists and their sentiments pervade the Declaration of Independence and subsequent expressions, they were deliberate in their determination not to establish a Christian nation with the mixing of God and government. Yet this was easier said than done as there are repeated references to the United States as God's chosen people, as the "New Israel," and words reflecting special or "manifest" destiny have been frequently recited at U.S. sacred ceremonies, and especially at the inaugural addresses of our presidents (Cherry 1998). Although the "great awakening" was a spiritual event in U.S. history which led to the evangelization of the countryside and the growth of the mainline churches, it was the Civil War which generated the depth of religion's political influence, and the subsequent influence on religion by political forces. Communities chose sides: mainline Protestantism was Republican, opposed slavery, and was centered in the North; Catholics and the poor of the South voted Democratic. The Southern Baptist Convention, mobilized during the Reconstruction, continued to identify with "the mythology of the Lost Cause" and thus saw themselves "like the wandering Israelites of the Old Testament, sworn to keep faith with the old way" (Phillips 2006, 145). This is seen as the birthplace of modern Christian fundamentalism and this group compiled "The Fundamentals" in 1910-1915 which was the blueprint for the radical right in the future.

Many factors account for the reversal of the political parties in the twentieth century where northern and urban areas converted to the Democratic ledger and the Deep South and Midwest became Republican strongholds. Mark Noll and Kevin Phillips, cited above, provide masterful studies of this phenomenon, detailing events which resulted in the Democrats clustering around political positions on war, society and civil disobedience that were too liberal for congregations who then moved to the Republican right. And thus Delaware, Ohio, like much of the United States, finds itself riding the crest of waves where religion is shaped by politics, economics and psychology. It is a community divided by categories of faith and religious practice.

How do we define those categories? Consensus would produce the following definitions.

Two Wings of Discipleship

These are the people who are likely to be in a house of worship on a given Sunday, who meditate on their existence and their faith from their specific theological heritage. These are the ones who energetically profess their convictions, who actively pursue illumination and sanctification, believing that profound commitment is required to deal with the challenges they must face.

This category is very heterogeneous. It is readily recognized in the contrasting wings of discipleship, the *traditional-to-liberal* groups on the left side and those now identified as more *conservative-to-evangelical* on the right. This nomenclature, of course, is less than satisfactory in that many on the left believe they are quite conservative in their passion to practice their faith, and many on the right take their traditions with great seriousness and sensitivity to social justice. Nevertheless they approach discipleship from very different perspectives.

The traditional-to-liberal group emphasize the importance of a genuine knowledge of their tradition which the theologian Douglas John Hall defines as "the ponderous weight of the past" (Hall 1993, 13), exploring all the foundations they have inherited from the church fathers to the great theologians, and expressing their faith in the great music and liturgies of tradition. They seek to search the inspired Scriptures as a source of illumination but, at the same time, a source which must be understood historically and contextually, as a commentary of the times. They seek not a set of dogmas or religious ideas and are dissatisfied with any "Sunday School theology" that is simplistic, that ignores the critical and disturbing doubts, anxieties and challenges of the modern world. Their commitment is to fully appropriate the past and to anticipate God's activity in the unfolding future of the world, including the empirical discoveries of science. Hall states his position simply: "I profess the faith as I remember its foundations and hope for its continuing illumination of the present and the emergent future" (1993, 16).

The conservative-to-evangelical group emphasizes that salvation does not come by intellectual knowledge and comprehension but by "grace alone." These persons seek a most personal experience with God, the *sine qua non* of sanctification, and to be known for their love of others, even as they are loved. As descendents of pietism they are suspicious of too much worldly involvement and insist that the faithful must profoundly repent and turn away from all things sinful if they are to be "saved" from damnation. They believe that the Bible, especially the New Testament, is, in essence, the inspired Word of God whose scriptures are to be searched "continuously." Doctrines and articles of faith are generally held as absolute and they are skeptical of claims that God's personal revelation has occurred in other religious leaders of history or in secular science.

There is, generally, a friendly tension between these two wings of discipleship. They can join hands in certain gatherings of good will, each believing that the other means well and that they share "a common life in Christ." But their challenges are earnest and sincere. Those to the left accuse those on the right of being thoughtless, swept up in an intellectual and theological vacuum which ignores modern Biblical research, which pays no attention to the evolution of Christian thought and historical theology, and which refuses to acknowledge the realities of basic science. Further the right is accused of being so obsessed with personal religious experience, or being "born again," that the result is a simplistic narcissism which is not linked to respected theological traditions. The right, in response, accuses those on the left of being contemporary Gnostics, seeking salvation by way of personal knowledge and without regard to personal commitment and moral discipline. They contend that those on the left have fashioned a way of discipleship which is easy, characterized by ethical relativism and no sacrifice.

Of course there are contradictions and inconsistencies in each wing. While those on the left criticize the right as being intellectually ignorant of Biblical theology, it is true that the vast majority of the membership on the left are almost Biblically illiterate in that they have little knowledge of such matters as the multiple authorship of the Pentateuch or of the difference between Pauline and Johanine theological traditions; and, of course, those on the right so faithfully attend to the study of scripture that they can often quote chapter and verse in a manner which mystifies those on the left. On the other hand, while those on the right lay claim to a personal and sacrificial commitment to discipleship, this has often been seen as a personal pietism which does not address the great social fractures all around us. The right may give handouts to the poor, yet the leadership for civil rights, social justice, and economic parity, has come from the left.

The Gathering Storm: Christian Nationalism vs. Christian Secularism

Churchill's metaphor of "The Gathering Storm" is not an exaggeration at this time in history. The dialectical tension between the right and the left in the disciple community is quite tolerable when contrasted with the more radical polarization that is occurring as each wing moves to extremes. To the far right is fundamentalism, and it represents the fastest growing segment of the U.S. religious scene. Fundamentalism is rooted in the belief of Biblical inerrancy, and its authority is absolute: anything which stands contrary to scripture is wrong and sinful. It proclaims imminent cataclysms and the approaching end of times possibly preceded by holy wars. Salvation is offered to those who repent and turn away from all things sinful, but it will be denied those who do otherwise. The study of theology is often considered tedious and tangential, a spiritual weakness which is certainly not of concern to the person in the pew. What is required is to "simply believe," to avoid all traps of ambiguity.

This radical development has produced a reaction on the left, the liberalism which sees the far right as nothing less than fanatical. The fundamentalists are viewed as being dogmatic and closed-minded in their total opposition to such issues as acceptance of homosexuals, abortion, and stem cell research. Their claims of having exclusive access to the truth are seen as dangerously arrogant and the pyrotechnics of their television ministries are offered as evidence of a primitive emotionalism that invites chaos. The far left accuses the far right of being incapable of tolerating ambiguity and assert that they resolve their anxiety by pledging allegiance to any group or personality who will offer authoritative answers and "relieve persons of their burdensome freedom" (Hall 1993, 26.) These charges, of course, invite the response from the right that has made the word "liberal" now tantamount to immoral, spineless, and godless. The positions are well drawn and will continue in the years to come. On the far left are those who now claim that they must now dissociate from any form of Christianity whose God is so hateful and instead promote a thoroughgoing secularism. On the right are those who see such a liberal as being evil, the tool of Satan.

Such polarization of extremes is hardly new in the history of religion, including Christianity. Schisms have led to eras of hatefulness, oppression, violence, and terrible persecutions. To the new student of history this is always perplexing: how could such suffering be precipitated by persons in pursuit of the sacred and averse to the profane? The solution lies somewhere in understanding the evolution of religion from social circumstances which tie the community to the spread of individual experience and personal autonomy (Durkheim 1961). It is certainly the singular characteristic of Christianity in North America since 1960 to become much more an expression of individual autonomy and much less a collective force. Most religious polls have reported the "detribalization" or the decline of membership of the various mainline denominations, and some even report the trend to no longer mention their denominational affiliation (Heinen 2005). But it is precisely the loss of the community's unified and shared sense of the sacred that provokes the polarization with each faction claiming ownership of the sacred and identifying those who disagree as profane or defiled. The subsequent hostilities now resemble ideological holy wars.

However, wars require armies, or, at least, political and economic power. Secular powers sometimes require the sanction, if not the blessing, of the religious authorities if they are to survive. The evolution of such a religious-political coalition in North America is very much a factor in any futurist assessment. It is fueled not just by demagogues or tyrants (although some on both sides are present) but also by psychological and political conditions in the United States which creates a vacuum that attracts religion. The psychological factor is the enormous anxiety being experienced across the United States, anxiety prompted by the realization that that there are no simple answers to problems of vast scope and incredible complexity: problems regarding the economy, the environment, terrorism and international policy, education, health care, and human relationships. Because there are no simple answers to these perplexing situations, persons are left with increasing ambiguity in almost every area of life. No society appears to be able to tolerate unrelenting ambiguity. It will move to resolve the anxiety with exclusive thought, adopting positions of "either black or white", of absolute right or absolute wrong. And the political factor quickly follows. The "far right" or the "far left" have an immediate appeal to a constituency rallying around absolute principles. The 2004 Presidential Election in the United States was characterized by "God and country" symbols, with "hot" issues such as homosexuality, abortion, stem cell research, and prayer in the public school inviting persons to choose between ambiguous process and fundamental truth. The voting patterns revealed that the cities were the more liberal constituency, struggling with ambiguity and affirming social changes, while the countryside was the more conservative, deeply convinced that a return to basic, time-tested, fundamental beliefs and practices was essential.

The coalition between the administration in power in the United States in 2004 and the religious right has been a matter of intense analysis and debate. Some would argue that the incumbent would not have been re-elected had it not been for the proactive support of the religious right which viewed him as the more "godly" candidate. Commentaries are blunt on the matter.

> The last two presidential elections mark the transformation of the GOP into the first religious party in history. . . . (We have) an elected leader who believes himself in some ways to speak for God, a ruling political party that represents religious true believers and seeks to mobilize the churches, the conviction that government . . . should be guided by religion. (Philips 2006, vii-ix)

And:

> One of the biggest changes in politics in my lifetime is that the delusional is no longer marginal. It has come in from the fringe, to sit in the seat of power in the Oval Office and in Congress. For the first time in our history , ideology and theology hold a monopoly of power in Washington. (Moyers 2004, xv)

This, of course, is the description of a formation of theocracy, a political system ruled by religion. An actual theocracy has been an anathema in the history of the United States, and images of the repressive Taliban in the Middle East have caused U.S. citizens to shudder. Yet many of the references in the 2005 State of the Union address linked freedom and liberty with divine wishes, and there was a clear theological version of Manifest Destiny, that this nation was called (by God) to democratize the Middle East.

Activists on both sides of this religious-political culture view the near future as becoming an increasingly intense theological soul war. The fundamentalists have named their movement Christian Nationalism portrayed bluntly by Pentecostal Pat Robertson: "There is no way that government can operate successfully unless led by godly men and women under the laws of the God of Jacob" (Philips 2006, 215). Or as the T.V. evangelist James Kennedy says:

> Our job is to reclaim America for Christ, whatever the cost. As the vice regents of God, we are to exercise godly dominion and influence over our neighborhoods, our schools, our government, our literature and arts, our sports arenas, our entertainment media, our news media, our scientific endeavors—in short, over every aspect and institution of human society. (Kennedy 2006, 1)

This group can be expected to mobilize its efforts for greater politicization of the churches in forthcoming state and national elections, proclaiming that the United States is itself profoundly sinful and must, as a nation, repent and renounce all things sinful. This, of course, includes voting down the "hot" social issues of abortion, gay marriage, stem cell research, and any form of turning away from the holy war against Islamic terrorists. Against this theocratic movement will emerge a coalition of the religious left and the Constitutionalists who will need to organize an infrastructure to engage those same churches with an appeal that religious persons learn to think for themselves. They will be supported by an increasing number of political and theological scholars who will, as Karl Barth did in 1934 Germany, sound an alarm

> against every form of groupthink, especially when those who claim an affiliation with God begin to rally hearts and minds under a nationalistic banner. When this occurs, it is the prophetic consciousness that brings a redemptive skepticism to our mythic realities. The confessing

> community can speak candidly and demand candor when career politicians and celebrity pundits feel trapped by a public that seems to only want promises of absolute victory in the war on evil, further proclamations of America's moral superiority, and assurance concerning the innate goodness of the American people. (Dark 2005, 155)

The Roman Catholic Experience

Catholicism has been affected by all the cultural developments described above. However, unlike Protestants, Catholics have no "competitors" for their membership. Further, Catholic dogma is so enriched with long history and tradition that it is not easily susceptible to shifting trends or radical developments on the left or on the right. And whereas actual church attendance and participation has declined, as reflected in all statistical studies, Roman Catholics still consider themselves full members of the Roman Catholic Church, even if inactive. If any religious group can anticipate a more steady state in the unfolding future, it would be this one.

There are also distinct differences and problems facing Catholicism in North America. As the new century begins, the Church continues to assimilate and be formed by decisions made some forty years ago at the Ecumenical Council known as Vatican II. Debate continues between the more liberal wing of the Church, which has enthusiastically embraced the changes introduced, and the more conservative, which believes the Council strayed too far from orthodoxy. However, commanding much more attention are events and developments which have absorbed the Church. The first, frequently called the greatest crisis of the Roman Catholic Church in the United States, has been the revelation of widespread sexual abuse—a betrayal of trust and the cause of great suffering and financial disaster. To its credit, the Church has been forthcoming with open analysis and the creation of national policies to prevent recurrence. The second is a critical shortage of priests. In the United States over the last thirty years. there has been a fifty-nine pecent decrease in the ratio of Catholic priests to that church's membership. Further, the existing priesthood is growing old—with an average age of fifty-nine—while the number of young seminarians is radically shrinking (Himes 2004). This has necessitated the emergence of a trained laity to assume leadership in parishes and to provide pastoral care. Church women are likely to become much more prominent in parish ministry.

It has been frequently noted that the millstones of the Roman Catholic Church grind "exceeding fine" but very slowly. Over 2000 years the Church has become quite familiar with upheaval and dispute. *Ecclesia semper reformanda est*; the Church is always in need of reform. Scholars are not so quick to forecast great changes in Catholicism and are suspicious of current trends. Further, they are reminded that the United States is home for only six percent of world Catholics and if there are to be major changes, it is more likely to occur in the Global South where there is little support for celibate clergy and where canon law is viewed as a European invention and readily challenged.

The Situation in Europe

The theological giants of the last century came from Europe. Their influence continues to be far-reaching. How strange, then, to commonly read of the decline of religion in Europe where only fifteen percent of the population considers it important to their lives—compared to fifty-three percent in the United States. The percentage of persons who attend religious services once a week varies: eighty-four in Ireland, fifty-five in Poland, forty-five in Italy, thirty-five in the Netherlands, twenty-five in Spain, twenty in France, and fourfteen in West Germany (Swanbrow 1997).

There would appear to be many factors at work that account for this phenomenon. Foremost is the existential reality that this is a population which has survived the horrors of two world wars, has lived at the center of the cold war, and has coped first hand with Marxism. Second is the problem of aging: the rapid growth in the older population and rapid shrinking of the younger. It is estimated that by 2030 people over sixty-five in Germany, the world's third-largest economy, will account for almost half of the adult population, and the nation will need to import one million immigrants each year to maintain its workforce. Many of those immigrants will be non-Christian. A third factor is the general character of Christendom in Europe. Although it arose from the intense doctrinal debates of the Reformation—debates which created competitive denominations in the United States—in Europe it produced Constantinian Christianity. In it, one is a Christian by virtue of one's baptism and birthright, whether or not one is religiously active. Indeed, dissent from organized religion may even be valued as a feature of independent thought.

However, any statistical picture of Europe as totally secular is misleading. There are many "new religious movements," sects and cults and a general religiosity, that appear to be growing rapidly. Moreover, there is the indigenous Pentecostal movement whose numbers are not easily counted but who are projecting an extraordinary increase in membership. This movement employs well-developed global networks and makes use of advanced communication technology, particularly in economically deprived countries (Inrovigne 2001).

The Situation in the Global South

If the Christian population has declined, or at least stagnated, in Europe, and if it is too often polarized by distrust and contempt in the United States, the situation in the Global South— both South America and Africa—stands in dramatic contrast. In that area the Christian Church is expanding in an explosion of vitality. By 2025, Africa and Latin America will represent the most heavily populated Christian nations. There a steady rate of growth during the last century has seen the number of Christians in Africa increased from 10 million to over 360 million. Africans and Asians now represent thirty percent of all Christians. Two-thirds of the Roman Catholic Church reside in Africa, Asia and Latin America and by 2025 that ratio should increase to seventy-five percent. By the middle of the present century it is estimated that there will be three billion Christians in the world, of whom one-fifth or fewer will be non-Hispanic whites. In contrast to Europe, half of the planet is now under age twenty-four with ninety percent of that cohort living in the Global South (Barrett, Kurian, and Johnson 2001).

These statistics are undoubtedly startling to most Western observers, as this period of expansion has not attracted attention. Perhaps the main reason is that the Christian Church in the Global South is not divided by the intense conflicts of the churches in the Global North. Philip Jenkins, in an arresting new study, analyzes the unique character of this "new Christianity" (2006). Instead of being defined by the doctrinal disputes that resulted from the European Reformation, or the "high" Eucharistic theology which was the Catholic reaction to Protestants, in the Global South there is a religious pluralism which accommodates diverse opinions and traditions including a more amicable interface with non-Christian religions who worship "the same God." The divisive issues of North America simply do not matter as much in the south. There is no "fundamentalism" which excludes or demonizes persons of different persuasion. There is no obstinate resistance to scientific fact, no wholesale discrimination against women or the sexually unorthodox.

What really matters to these Christians, what excites and unites them, is a profound respect for the authority of the Holy Scripture and a special—and quite natural to the culture—interest in its supernatural elements: miracles, vision, healings, and prophecy. These people do not fall into the trap of Biblical inerrancy; they are open to interpretations enhanced by modern scholarship. They do, however, take each word of the Bible very seriously, believing that if it is studied, it will inspire and reveal strategies to address poverty and social injustice.

Jenkins' study reveals Southern Christian churches as dynamic and animated. Their members are characterized not by doubt, anger, and competing claims to faith, but by an excitement in the study of the Bible, both its Old and New Testaments. Such Biblical settings as pastures, meadows, mountains, ancient lands, far and distant countries—and events involving shepherds, tyrants, plague, and famine, slavery, prophets, and blood sacrifice—are all familiar to Africans. They take ownership of the Biblical account of the search for true wisdom, for the moral law, blessings for the poor, magic, healing and the cosmic struggle between good and evil. "It is precisely the 'primitive' features of ancient Hebrew religion, which distress many modern westerners that have long endeared it to Africans" (Jenkins 2006, 47). This is also true in Central America.

> Guatemala certainly feels Biblical. Sheep, swine, donkeys, serpents—these are everywhere as are centurions, all manner of wandering prophets, Pharisees, lepers. It is the perfect backdrop for religious parables about fields, both barren and fertile, hunger and plenty. . . . Millions know roads where a traveler is likely to be robbed and left for dead . . . (or) they understand that a poor woman who loses a tiny sum of money would search frantically for coins that could allow her children to eat that night. . . . (they) appreciate the picture of the capricious rich man who offers hospitality on one occasion but on another day demands payment of exorbitant debts . . . the corrupt official of a ruling party. (Jenkins 2006, 69)

Islam and Judaism

This study has been devoted to Christianity which is presumably the culture in which most readers find their questions and concerns. The scope of the study does not permit any extensive analysis of other world religions and their future. However two of those religions have something in common with Christianity at this juncture. Both Islam and Judaism are experiencing the same internal dialectic as Christianity, namely the conflict between those who try to modernize the faith and those who fear such modernization and thus retreat to traditional orthodoxy and fundamentalism; and, with this, the interaction of politics and religion, with each shaping the other, resulting in the potential polarization of theocracy and secularism.

Islam

Islam is the second largest religion in the world, with approximately half of the membership that Christians report, and constituting around twenty percent of the world's population. It is growing rapidly whereas the worldwide Christian population is declining. Muslims in the United States are estimated to number around 2.8 million; this is more than the number of U.S. Presbyterians. One in four Muslims in the United States is African-American (Tishler 2007, 358).

It is likely that if a survey were taken even twenty years ago, most Americans would have demonstrated little if any understanding of Islam, viewing it as an ancient monotheistic religion where God is called Allah, and Mohammed instead of Jesus is seen as the central religious figure. In higher education, students were taught of the origin of Islam in the seventh century, that the Qu'ran is its sacred literature, and that Muslim families, with socially conservative values, often find it difficult to reconcile the "American way of life" with their traditional moral values. Some would have learned that there are more radical fundamentalist groups, having noted the influence of the Nation of Islam in the United States and the Taliban in Afghanistan.

September 11, 2001 changed that perception. Suddenly the world was aware of a religious people with members who had long hated the ideology of the West, seen as the ancient enemy of Islam, and especially hated the United States which was blamed as the source of excessive modernization, shameful values, and also a supporter of arrogant colonialism. Then with the ensuing war in Iraq, Americans became familiar not only with the fierce rivalry between two of the Muslim divisions, the Sunnis and the Shiites, but also with how both factions appeared to reject the United States' attempt to democratize the Middle East.

It is likely that these dramatic events have led to a common misperception of Islam where a very small theocratic minority has seized the stage and become empowered by the Western response to its radical voice. It is important to know that the current rejection of modernity as a core ideology can be traced to Muhammad ibn 'Abd al-Wahhab in the eighteenth century who viewed Western values and the advance of Christianity as corrupting and debasing the true Islamic heritage. Wahhabism became a major force in the entire world only when the Saudi kingdom was consolidated in the early twentieth century. All Western nations coveted the vast reserves of oil in Arabia and consequently supported and enriched the Saudi regime. Wahhabism became the state-enforced doctrine of one of the most influential governments in all Islam, which disseminated its doctrines and funded its propagation (Lewis, 2003).

Among the vast majority Muslims, however, mainstream Islamic doctrine was followed which welcomed change, modernization, and peace with Christianity. But political events continued to occur which magnified tensions and fed polarization. The fate of the Palestinians in their conflict with Israel was a daily story of frustration and humiliation. Then the failure of both capitalism and socialism in the Middle East fed the idea that Western ideas are the ultimate source of misery in the Islamic world. Wahhabism fed on these events and impeded the efforts of those who advocated modernization of their nations.

One Middle-Eastern scholar in addressing the future of Islam has driven home the point that the current upheaval in the Middle East is not, as popularly construed, the moment of the clash of civilizations between the West and a pan-Islamic Jihad, but rather a moment in a fourteen-century conflict within Islam. He argues that many Muslim leaders around the world desire some of the Western principles of enlightenment: human rights, individualism, constitutionalism, and the rule of law. Fundamentalists accuse these reformers of promoting colonial oppression. The reformation appeared to be gaining ground, moving toward areas of Islamic democracy when the fundamentalists did the one thing they knew would galvanize support: a spectacular attack on United States soil which would prompt the United States to overreact. He writes:

> After September 11, it went exactly as they had planned. We not only gave these small groups a disproportionate amount of attention, but also a disproportionate amount of power. . . . That is the whole point of terrorism—it's the tactic of the weak. Its purpose is to give the

> illusion of power. . . . We also handed them the language with which to transform the so-called war on terrorism into a war against Islam, a war against Muslim values—another colonialism endeavor, or in other words, another Crusade. (Chauddrhy 2006, 2)

Another scholar states that the challenge of Islam in the new century is to develop a theology of reform and address Islamic extremism. John Esposito of Georgetown University says that as a result of terrorist activities, from the Iranian revolution through 9/11, "Americans and Europeans tend to hold a monolithic view of Islam rather than understand its broad diversity as a religion found in 56 countries" (Esposito 2003, 1). As a result many persons believe that Islam is incompatible with democracy, and Esposito reminds us that most Muslim nation-states were formed only after World War II, have artificially drawn boundaries, and that their leaders are kings who rely on armed might, not popularity, to remain in power. Many Muslims oppose extremism as well as American policy, but they also oppose their own governments and want to change the system from within.

> Unless we realize that distinction, when we look at political Islam we're not going to understand what's going on in the Muslim world for the next 20 to 30 years. . . . There is no country in the Muslim world that's not going to have religion in some way visible in state and society. (Esposito 2003, 11)

Chauddhry predicts that the near future of Islam is likely to be a very bloody and violent chapter of politico-religious history. He is, however, very hopeful that the pendulum will swing, that Muslims will all see that extremists are contributing to their destruction, and that this is not the faith of Islam. He warns the West that while it should promote democratic values of human rights, social and economic progress, it is imperative to allow the Muslim world to shape its own form of democracy, including forms of government which do not insist on a strict separation of church and state. He reminds the United States that it is not strictly secular, that its constitution and many of its laws, social morals and values derive from Judeo-Christian religion.

Judaism

At first one would not expect to list Judaism as a religion to be effected by the unfolding events of the future. For if any religion is considered stable, constant and not given to vicissitudes of current events, it is Judaism. Until the eighteenth century there was basically only one kind of Judaism, now called the Orthodox. However, due to the Enlightenment and Emancipation, certain reform movements began which allowed for major reinterpretations of religious practices and customs.

Then came the Holocaust when millions of Jews, including a large majority of the Orthodox, perished. This was followed by the creation of the state of Israel which radically reshaped the Jewish experience. Many Jews identified with Israel as a secular replacement for religion and found there deep psychological and emotional succor, a homeland of safety in a world history of persecution.

This, of course, catapulted Judaism into the very center of world conflict. The state of Israel is an island at the center of an Islamic ocean, the very definition of instability. The long-standing, unresolved Palestinian-Israeli conflict only serves to complicate matters. On the other hand, Israel's accomplished peace with Egypt, the largest and most powerful Arab state, and peace with the Kingdom of Jordan, provides a degree of hope and stability.

Even though it is the Orthodox population of Jewish community that is numerically increasing in the United States, the non-Orthodox religious movements of Reform, Conservative and Reconstructionism, representing the majority of affiliated American Jews, continues to be dynamic and creative.

Future Trends—2020

From this study it is most reasonable to infer from current developments the following understanding for future trends.

- Religion and politics will continue their intense courtship. In the United States there will be no simple solutions to the complex interaction of social, economic, political and environmental problems. The resulting ambiguity will be unacceptable to many who, in their anxiety, will require direct, simple and authoritative or "black-white" equations. This will fuel an increased coalition between the far right of both religion and politics and will be featured in proclamations from both the pulpit and the campaign trails. Each faction will believe that the coalition is essential to solidify a base of power and influence. In response to this the far left will seek to expose the implicit threat of theocracy and will attack fundamentalism in both religion and politics while promoting a "Christian secularism." This conflict will further discredit mainstream religion, itself already in continuing decline.

- The Roman Catholic Church in North America will not be easily influenced by the politico-religious coalitions but will be fully occupied with internal problems, notably the diminishing number of priests. It will experience the emergence of new lay leadership which will command increasing authority.
- Europe will experience a continuing decline of organized religion, replaced by the "Disestablished," i.e., persons finding religious faith more through individual experience than collectively. However indigenous sects, cults and Pentecostal groups will experience significant growth.
- Religion in the Global South will continue to expand at an exponential rate and will become, for all practical purposes, the very matrix of Christianity. The consequences of this are vast including major challenges to canon law, the ending of the requirement of celibacy for priesthood, and the genuine possibility of election of a Pontiff from that hemisphere.
- As long as Western powers intervene in the Mid-East with the stated mission of democratizing the land, the extreme Islamic fundamentalists will be empowered and legitimatized in their terrorist strategies, seeking to completely eliminate Western influence. If, however, the conflicts within Islam are understood as internal matters, and if moderate Islam is supported in its effort to promote theological and political reform on its own terms, and if it is accepted that there will be a clear relationship between Islamic doctrine and government, although not a theocracy, then co-existence with the West will be implemented and the "holy wars" can cease.
- Judaism and the Jewish people continue to be a major player in the Middle East. The complex Middle East conflict will continue to be a source of global concern even as significant strides are being made toward resolving the conflict.

References

Adler, Jerry 2005. "The Search of The Spiritual." *Newsweek,* September 5, 2005.

Barrett, David B, George T. Kurian and Todd M. Johnson. 2001. *World Christian Encyclopedia,* 2nd ed. New York: Oxford University Press.

Chauddhry, Lakshmi. 2006. "The Future of Islam." http://www.alternet.org/war on iraq/21891/. (accessed June, 2006).

Cherry, Conrad. 1998. *God's New Israel: Religious Interpretations of American Destiny*. Chapel Hill: University of North Carolina Press.

Dark, David. 2005. *The Gospel According to America.* Louisville: Westminster John Knox Press.

Durkheim, Emile. 1961. *The Elementary Forms of Religious Life.* New York: Collier.

Esposito, John. "A Hard Look at the Future of Islam." http://news-service.stanford.edu/news/2003/november19/islam-1119.html. (accessed June, 2006).

Hall, Douglas John. 1993. *Professing the Faith: Christian Theology in a North American Context.* Minneapolis: Fortress Press.

Hammond, Phillip. 1992. *Religion and Personal Autonomy: The Third Disestablishment of Religion in the United States.* Columbia: The University of South Carolina Press.

Heinen, Tom. "Churches, Just Without the Label. Seeking Outsiders Congregations Drop Denomination—or More—From Their Names." *Milwaukee Sentinel*, August 2005. http://www.washingtonpost.com/wp-dyn/content/article/2005/08/05/AR2005080501525.html. (accessed June, 2006).

Himes, Michael J., ed. 2004. *The Catholic Church in the 21st Century.* Liguori, Missouri: Liguori Press.

Inrovigne, Massimo. 2001. "The Future of Religion and the Future of New Religions." http://www.cesnur.org/2001/mi_june03.htm. (accessed June, 2006).

Jenkins, Philip. 2006. *The New Faces of Christianity: Believing the Bible in the Global South.* New York: Oxford University Press.

Kennedy, James. 2006. http://www.theocracywatch.org. (accessed June, 2006).

Lewis, Bernard. 2003. *The Crisis of Islam: Holy War and Unholy Terror.* New York: Random House.

Moyers, William. 2004. Speech to the Center for Health and Global Environment, The Harvard Medical School, December 4, 2004. Quoted from Philips, *American Theocracy*. New York: Viking.

Noll, Mark. 2002. *The Old Religion in a New World.* Grand Rapids: Eerdmans.

Olson, David. "Empty Pews, Signs of Hope." *The Covenant Companion,* February, 2006.

Philips, Kevin. 2006. *American Theocracy.* New York: Viking.

Pressor, S. and L. Stinson. 1989. Data Collection Mode and Social Desirability Bias in Self-Reported Religious Attendance. *American Sociological Review.*

Smith, Huston 2001. "Religion in the Twenty-First Century". http://www.vedanta.org/reading/monthlyarticles/2001/11.21st_century.html. (accessed June, 2006).

Swanbrow, Diane. 1997. "Study of Worldwide Rates of Religiosity, Church, Attendance." University of Michigan News Release. http://www.umich.edu/-newsinfor/Releases/1997/. (accessed June 2006).

Tillich, Paul. 1966. *The Future of Religions*. Ed. Jerald C. Brauer. New York: Harper and Row.

Tishler, Henry L. 2007. *Introduction to Sociology*: 9th Edition. Belmont, CA: Wadsworth/Thomas Learning.

Chapter 3

Education

Suellen Mazurowski, JD

> These are the things I learned:
> Share everything.
> Play fair.
> Don't hit people.
> Put things back where you found them.
> Clean up your own mess.
> Don't take things that aren't yours.
> Say you're sorry when you hurt someone.
> Wash your hands before you eat.
> Flush.
> Warm cookies and cold milk are good for you.
> Live a balanced life. . . .
> Take a nap every afternoon.
> When you go out into the world . . . stick together.
> Be aware of wonder.
>
> (Fulghum 1988, 4-5)

According to India's Swami Krishnananda, the real purpose of education is to draw out the best and noblest of the latent faculties in humanity. (Krishnananda 2006). Krishnananda claims that moral achievement is an integral part of education. He states that it is imperative at least one lesson in a week, if not every day, should be taught on morality and spirituality. He finds fault with the current system of education in

India, stating that it was introduced by alien rulers who sought to "enable the unsuspecting Indian to qualify himself to be able to serve the ruler" (2006, 1).

The American system of education may also be designed to "serve the ruler." In his book *Tough Love for Schools,* Frederick M. Hess states:

> From the time of Plato's "republic" to Dewey's "lab school," philosophers have understood that society could use schooling to forge ideal citizens. In the contemporary American context, there resides the hope that schools can shape skilled, responsible, and self-sufficient citizens, strengthening the nation and alleviating any further need for government assistance (Hess 2006, 17).

From the above it appears that government plays a large role in shaping the process of education. If so, one might question whether education itself is serving its true function. If Krishnananda is to be believed, the true function of education is to draw out the best faculties in mankind. However, what we actually hear today is that our youth are not achieving their academic potential and are not therefore "serving the ruler."

Reporter Kathryn Wallace has complained that students in the United States are falling behind those of other countries in math, science, and engineering. Thus America is steadily losing its global economic edge (Wallace 2006). It appears that China, Japan, India, and other countries are challenging United States dominance. Our failure to be competitive in science will widely affect our standard of living, national security, and way of life (Wallace 2006).

During the last century our country has moved from an agricultural economy to a manufacturing economy and finally to a technical and information-based economy. Math and science are the tools of a technical age. Cries for reformation of education are being heard nationwide. As citizens of the United States, we must decide what changes, if any, should be made.

Historical Perspective

In early civilizations adults taught their children the knowledge and skills they would need to master and eventually pass along to the next generation. Instruction was given orally, often through story telling. Formal schooling began when cultures began to expand their knowledge beyond the minimum required for survival. The earliest formal schools began many centuries before the birth of Christ (Cheeseman 2006). Until the end of the Middle Ages most education was provided to the sons of the wealthy and noble; therefore only a few people could read and write.

In the Western world formal early education was provided mainly by priests and prophets. In the Middle Ages the Roman Catholic Church took charge of most education in areas under their influence. Many of the first universities in Europe have Catholic roots. There were however some secular universities established, such as the University of Bologna. Books were rare in ancient times because they had to be copied by hand. After Johannes Gutenberg invented the printing press around 1440, books became more widely available. The availability of more books revolutionized education and gradually raised the literacy rate of the general public (Cheeseman 2006).

The first public school in America was established in 1635 in Boston, Massachusetts by the Pilgrims (Cheeseman 2006). Early American schools focused on religion because it was widely believed that it was essential for a student to learn to read the Bible. At the beginning of the nineteenth century, the industrial revolution brought a demand for workers who possessed at least minimum literacy. The need for literate workers caused governments to begin mandating attendance at standardized schools with a state-prescribed curriculum. During the twentieth century children began spending longer periods in formal education before entering the workforce (2006).

Recent History of Education in the United States

An examination of the recent history of education in the United States shows us that politics has greatly influenced the field. For more than fifty years Democrats and Republicans have struggled over the role of the federal government in education and the issues of federal funding and control of education. Frederick Hess has dedicated the first chapter of his book, *Tough Love for Schools,* to giving us an expansive review of the politics of education in the United States (Hess 2006). A brief summary of this review follows.

There was little need for extensive education of the populace when our republic was founded. The nation had an agricultural economy. The federal government played no role in education until 1867 when the United States Office of Education was established. In 1930 less than one-fifth of adults over twenty-five years of age had completed high school. With increasing industrialization and the passage of the GI Bill after World War II, high school graduation became the norm and college enrollment soared.

The 1954 U.S. Supreme Court decision of 1954 desegregation case, *Brown v. Board of Education* gave birth to the public idea that a good education is a birthright in a free society. In *Brown,* the court emphasized the importance of equal opportunity for all citizens, both black and white.

Russia launched Sputnik, the first space satellite, in 1957. Fear that the United States was lagging far behind in math and science led to the passage of the 1958 National Defense Education Act (NDEA). The NDEA significantly increased federal funding for education. The 1965 Elementary and Secondary Education Act (ESEA) was passed under President Lyndon B. Johnson. It was the first truly comprehensive package of federal aid to education. However, even after the ESEA, the federal government still contributed less than ten percent of total education spending.

By 1970, more than seventy-five percent of young adults in the country had finished high school and more than sixteen per cent had completed college. Increasingly Americans were seeing education as a way to advance economically and socially. Democrat Jimmy Carter, who was widely supported by the nation's public school teachers, established a cabinet level Department of Education position in 1979. By 1989 the Democrats had established themselves as the party of education, largely by supporting school teachers and advocating steady increases in federal education spending.

In the1980 election, Republican Ronald Reagan called for school vouchers that would permit public funds to support parochial school tuition. He further called for slashing taxes and government spending, ending federal government regulation of public education, and elimination of the federal Department of Education. In 1981 Reagan reformed much of the Elementary and Secondary Education Act instituted by Eisenhower when he won passage of the Education Consolidation and Improvement Act (ECIA). The ECIA reduced federal funding for education by twenty percent and increased flexibility in the states' use of federal funds.

Reagan also named a high profile commission to produce a report on the state of American education. The report, *A Nation at Risk,* warned that the failure of the United States to keep pace with the educational system of Japan was putting our future in jeopardy (National Commission on Excellence in Education 1983). *A Nation at Risk* put education on the national agenda and led to the idea that the federal government should require state and local educational authorities to meet minimum national standards.

In 1988 George H. Bush promised to be the "education president." Republicans continued to advocate for "school choice" as a way to neutralize the party's weakness on education and the fairness issue. Republicans argued that federal school vouchers would enable the children of low income families to attend better schools.

The campaign platform of Democrat William Clinton called for governments to provide equal educational opportunities for all. Meanwhile, before and after the 1992 elections, Republicans continued to advocate choice-based school reforms, such as public school choice, charter schooling, and the use of school vouchers. After the 1994 mid-term elections the Republicans also tried to cut spending for the federal Department of Education and to eliminate the department altogether.

In the 2000 election, George W. Bush campaigned on a platform of "compassionate conservatism," emphasizing education, and promising that "no child would be left behind." He advocated school choice, opposed significant increases in educational expenditures and put the blame for educational problems on school districts and teachers. One of his first acts in office was the passage of the *No Child Left Behind Act* (Hess 2006).

Current Issues in Education

The No Child Left Behind Act

One year after he took office President George W. Bush signed the *No Child Left Behind Act* (20 USC Section 7801 et. seq. 2002). *The No Child Left Behind Act* (NCLB) creates federal directives regarding test use and consequences, puts the federal government in charge of approving state standards and accountability plans and sets a nationwide timetable for boosting achievement. Sanctions are prescribed for underperforming schools, including loss of federal funds.

The NCLB requires that each child shall have a highly qualified teacher. It requires all public schools to annually test all their students in grades three through eight in reading and math and requires that every state measure whether its public schools are making "adequate yearly progress" (AYP) toward universal proficiency in those subjects. Science was added in 2007-2008. Each school must annually show steady improvement in every grade. Schools must also demonstrate improvement in educating various population groups including disabled, non-English speaking, and minority students. If a school fails for one year it is deemed "in need of improvement." If it fails for a second year the local district is required to provide each child in that school a choice of alternative public and charter schools that are making satisfactory progress. The NCLB provides federal funds to pay for after-school tutoring for students if a school fails for a third year. If a school fails for a fourth year it must write a school improvement plan. A school which fails for a fifth year must be "reconstituted." Federal regulations tell the states how to implement the NCLB. State education departments set standards, create tests, and intervene in districts that fail to meet Adequate Yearly Progress goals. The local school district has the responsibility to fulfill the largely unfunded federal mandates.

It is too early to determine whether the *No Child Left Behind Act* will accomplish its goals. The effectiveness and desirability of its measures are widely debated. On the political left, public education interest groups are pushing for an end to the testing and accountability programs. On the right, conservatives are calling for a return to local control. The strengths of the act are that it has forced schools to begin providing more information about student performance. Testing and accountability provisions have forced some schools to make decisions about how to generate improved performance. A major weakness of the act is that the passage rates on state tests are used by the federal government to determine whether the state meets federal standards. Many states, such as South Carolina, have standards which are higher than the standards set forth in the *No Child Left Behind Act*. Studies show, however, that some states are watering down their own achievement standards to avoid accountability sanctions. In at least twenty states, scores on state examinations have improved while children in the same states have not posted similar gains on the federally mandated National Assessment of Education Progress test. This has led some experts to declare that NCLB has started a "race to the bottom" in terms of lower state standards (Lips 2005).

While debates over the NCLB continue, students in our schools struggle with test anxiety and school districts spend many tax dollars to administer and grade examinations. Edna Crews, former Superintendent of the Beaufort County South Carolina School District, reported to this writer that the results of the Palmetto Achievement Challenge Test (PACT) used by South Carolina schools are of limited practical use to the teachers in the schools. Test results usually are reported to the school months after the examinations are taken and after the end of the school year. This is too late for a child's teacher to do anything to remedy any educational deficiencies of the child in the year in which the test was given (Crews 2008). Furthermore PACT tests give only a score. They fail to pinpoint specific areas of academic weakness. Ms. Crews was successful in implementing the "Measures of Academic Progress" (MAP) testing program in the Beaufort Schools. MAP tests are given to students three times a year. The tests are designed to be *progressive*; if a student does well in a subject area, questions in that area will become more difficult on the next test. MAP test results *are* scored in a timely manner and they *do* detail the specific areas in which a student is strong or weak. By using the MAP tests educators in Beaufort County are able to use the examinations as a teaching tool which is geared to individual student achievement. Each student can advance academically throughout the school year from the point where he or she was previously. Since this interview the school district has discontinued the use of PACT tests.

Testing is expensive. *Ohio Schools on Line* for April and May, 2006, reports that compliance with the NCLB will cost Ohio's school districts a total of $1.491 billion annually (Ohio Education Association Research Bulletin 2006). This sum represents an eleven percent increase over current total operating budgets. The web article reports that ninety-seven percent of the expected costs with the NCLB are unfunded and that additional federal funding is expected to cover only $44 million of the $1.5 billion in costs. The article concludes by stating that expecting the state to absorb the cost of NCLB is not realistic. The burden of funding NCLB is mostly placed on the local school district. However, Ohio voters are slow to approve new tax levies. The authors close by stating "[I]t can only be concluded that the rational decision may be to leave NCLB behind" (Ohio Education Association Research Bulletin 2006, 3).

Not only is testing expensive, but also the test results often are not useful as a tool to improve instruction, and extensive testing causes anxiety for teachers and students. At a minimum we might ensure that testing serves as a viable tool to implement academic progress. We also might inquire whether there is danger in having the schools mainly directed by test scores, rather than by what is appropriate and best for each individual child. What happens to the academically limited child whose strengths lie in other areas, such as music or art? In our haste to improve test scores are we failing to draw out the best faculties in each individual child and to help each child discover his or her place in the world?

Tax Vouchers

Many states, including Arizona, Pennsylvania, Florida, Wisconsin and Ohio have enacted tax credit voucher plans. Essentially, tax vouchers are publicly funded scholarships that permit public school students to attend another school of their choice. Vouchers are partially based upon the idea that providing a choice of schools to U.S. residents will place pressure upon the public schools by depriving them of a share of the funding that they receive for the pupils who leave the school, thereby forcing public schools to provide better education to their pupils. Voucher amounts and the requirements for student eligibility vary widely. A voucher may represent only a small percentage of the cost of attending the chosen school.

Many taxpayers have opposed school vouchers for attendance at parochial schools, arguing that the use of public funds for parochial schools constitutes an unconstitutional governmental aid to religion. However, the United States Supreme Court has held in at least one case that the use of school vouchers is constitutional if the purpose of the tuition aid is to improve the education of students and not to aid in the establishment of religion (*Zelman v. Simmons-Harris* 2002).

U.S. citizens have been divided over the issue of school vouchers for political reasons as well. Between forty-eight and sixty percent of African-Americans support school vouchers (Hess 2006). Scholarship aid to attend another school can remove a child from a substandard school building and place the child in a modern facility where he or she may receive more rigorous instruction from better teachers. However, middle class suburbanites who have sacrificed to purchase an expensive, heavily taxed home so that their children can attend a school in a good school district, often oppose the use of vouchers (Hess 2006).

While the voucher movement has gained some ground, currently only a small percentage of the nation's public school students utilize vouchers. There are insufficient voucher funds available to solve the problem of inadequate schools. Even if there were sufficient voucher funds there must actually be alternative schools for the children to attend. In commenting upon the failure of the South Carolina legislature to pass a recent voucher tax proposal initiative, Edna Crews, former Superintendent of the Beaufort County South Carolina School District, points out that voucher programs only work "when parents have real choices of schools, such as in urban areas with parochial schools of high quality." Such is not the case in Beaufort County, SC, and in many other areas at this time.

Magnet Schools

Magnet schools are public schools operated by a local school board that emphasize a specific subject area, teaching method, or service to students. Many people are familiar with the use of magnet schools as a teaching tool for the performing arts. Such schools are often part of a large urban school district. Since many charter schools are chartered for a specific purpose and many offer particularized themes, *Pittsburgh Post Gazette* writer Carmen J. Lee predicts that the number of magnet schools will decline as the number of charter schools increases (Lee 2000).

Charter Schools

In charter schooling the state permits an individual or a group to open and manage a state-funded school. These schools may be chartered by school districts, state school boards, churches, city governments, universities and other organizations. The charters of the schools can free them from many state regulations that apply to district schools. However, charter schools are not freed from accountability. Schools that fail to produce the results detailed in their charters or that fail to uphold the law are to be closed. Currently however, charter school accountability is primarily about closing schools with low enrollment, facility problems, financial improprieties or mismanagement, rather than about monitoring or ensuring academic performance. Charter schools currently enroll less than two percent of the K-12 pupils in the nation (Hess 2006).

A *New York Times* editorial tells us that the charter school movement began with the promise that independently operated schools would outperform their traditional counterparts if they could be exempted from state regulations while receiving public money (Collins 2006). Charter laws are now on the books in about forty states. However, several studies have shown that charter schools perform no better than other public schools. Moreover some states have opened so many charter programs so quickly that they can barely count them, let alone monitor student performance. Collins complains that some charter schools actually harm achievement. Where charter schools have failed, states often lack the political will, or even a process for closing them. Promising charter systems are few but those that exist have some things in common:

1. The states issue charters only after a rigorous screening process;
2. They provide technical assistance to the schools; and
3. They provide sophisticated oversight (Collins 2006).

It is suggested that states abandon the strategy of giving public money to private schools and then looking the other way (2006). Regardless of criticism, however, the bandwagon for charter schooling continues to roll on, with the successes of individual charter schools being detailed in the media.

Technology

Whereas the first part of the twentieth century saw a shift from an agricultural to a manufacturing society, its latter part saw a shift toward an increasingly technological society. At the birth of the twentieth century it was sufficient for a child to learn how to read, write and perform simple arithmetic. However, many of today's workers must be digitally literate to successfully find employment.

Students must be taught to use the new technology. At the same time that technology is itself being used in public schools and in homes across the nation to teach academic subject matter. Today, parents who wish to assist their children can walk into any Wal-Mart, Sam's Club, Staples, bookstore or library and choose from a wide variety of educational software. Recently North Carolina announced that an executive director has been hired to head up "NC Virtual," an on-line high school, which will enable North Carolina to offer a great many more advanced placement courses to its students (Billingsley 2006). The on-line school will also allow students to make up missed credits for graduation, work at their own pace and hone their time management skills.

The Microsoft website reports that Microsoft and the School District of Philadelphia have joined forces to create a high school that exploits innovation and technology (Microsoft 2006). Technology is incorporated in curriculum delivery, community collaboration, back-office support, content creation, and dissemination of content and assessment (2006). The school opened in the fall of 2006 for the first 170 students.

The school is a $63 million building which will eventually allow 750 digitally-enabled students to be served at home and at school through wireless broadband connectivity (Hogan 2006). The school's mission is to create a next-generation learning laboratory where new curricula and learning methods will be devised and tested. It will also be a research and development complex, where visiting educators can witness the action and then go home and duplicate it in their own districts. From the beginning the district has had access to Microsoft personnel and the company's corporate knowledge base to help build the school (2006). The design of the school of the future is forward-looking, yet the 168,000 square foot project has remained within a traditional budget (Cohen 2006). It borrowed Microsoft's organizational philosophy to make the most of available funds with pointers ranging from the purchase of construction materials to the hiring of staff.

Bill Gates has stated that harnessing a passion for technology and applying it to learning will empower people of all ages, both inside and outside the classroom, to learn more easily, enjoyably and successfully than ever before. Gates has further stated that improving education is the best investment we can make because down stream benefits flow to every part of society (Microsoft 2006, 5). The dedication of Microsoft to education and its participation in building the School of the Future is an impressive endeavor which will serve as a model for many future schools.

Philanthropy

Bill Gates has blessed us with his passion for technology. However he also has a passion for philanthropy (Microsoft 2007). He and his wife Melinda have established the Bill and Melinda Gates Foundation. In June, 2006, Bill Gates announced that after July, 2008, he would transition out of a day-to-day role in the Microsoft company to spend more time on his global health and education work at the foundation. The foundation has committed billions of dollars to improve learning opportunities through myriad projects, including the Gates Library Initiative to bring computers, internet access and training to public libraries in low income communities in the United States and Canada (2007). The foundation has also utilized a variety of creative methods to reach young people through the media, including showing documentaries on MTV, showing students the value of obtaining a high school diploma and providing on-line programs to present information to young people (Gates Foundation 2007). The scope of educational aid provided by the foundation is too vast to detail here. However the philanthropic efforts of this foundation will make it a major player in assuring educational progress in coming years.

Privatization

Private entities have long been involved in the educational process through such activities as the sale of educational materials and the provision of testing services. However, passage of the *No Child Left Behind Act* greatly expanded opportunities for private entities to profit from education.

The NCLB promises poor parents a free tutor for any child who attends a school that receives federal poverty aid but who has not made steady progress for three consecutive years. This section of the NCLB has provided an economic boon to some (Feller and Welsh-Huggins 2006). They report that a couple in Ohio has purchased a franchise that provides one-on-one academic help. Much of their business is tutoring, funded by the *No Child Left Behind Act*.

Revenues for products and services sold to public schools rose six percent in 2004-2005. Much of it is flowing to educational testing services. In the last five years Pearson Education Measurement's contract with the State of Texas has risen from $36 million to $60 million a year. Marketing firms are also profiting from the demand for data that describes how schools are falling short, with the information broken out by grade, subject, and names of teachers who instruct the pupils. Here are some revealing numbers:

1. In 2006 the government had 2.5 billion dollars available for schools to hire tutoring companies. However only 233,000 of 1.4 million eligible children took advantage of free tutoring in the 2003-2004 school year;
2. The revenue from products and services sold to public schools in 2004-2005 was $22 billion dollars; and
3. School districts were expected to pay $500 million for tutoring in 2006 (Feller and Welsh-Huggins 2006).

Private entities have also begun to actually run schools or partner with schools and school districts. Edison Schools, Inc., a for-profit education management organization for public schools in the United States and Canada, estimated that in the 2005-2006 school year it would serve approximately 330,000 public school students in twenty-five states, the District of Columbia and the United Kingdom (Edison 2006).

Edison also partners with charter schools. It claims to dramatically improve the education of children, increasing their scores on standardized tests, including those of students in economically disadvantaged districts. According to Edison its success comes from highly developed management systems, professional development programs for teachers and administrators, extensive use of technology, and efficient systems that allow more dollars to be allocated to the actual education program (Edison 2006).

The activities of Edison meet with mixed reviews. The *Toronto Globe and Mail* reported on October 30, 2002, that Edison, which had contracted with the Philadelphia Public Schools to manage twenty of the poorest-performing high schools, liquidated most of the textbooks, computers, lab supplies and musical instruments the company provides the day before school started, leaving students with decades-old books and no equipment. The paper reported that this action followed the summer stock market crash of 2002, when Edison, formerly a publicly traded company, saw its stock drop from the year's high of $21.68 per share to less than $1.00 on the NASDAQ Exchange (Saunders 2002).

The Public Broadcasting System has also reported extensively on Edison Schools. The PBS *Frontline* program of July 3, 2003, reported that the Wichita Board of Education entered into several contracts with Edison to operate schools. Initially the district was pleased with the results achieved by Edison. However, eventually the district terminated the contracts, concluding that it could achieve the same results as Edison more economically by implementing some of the Edison techniques. Baltimore, Maryland, however, reports excellent results in school improvement after contracting with Edison (Public Broadcasting System 2003).

A critical public issue raised by *Frontline* is whether for-profit companies should even be in the business of running public schools. PBS questions whether running private schools for profit runs the risk of putting goals of investors ahead of needs of students. *Frontline* also notes that the debate over privatization is political. The right argues that competition, innovation and the potential for private reward is precisely what is needed in education. The left fears that the one institution that is untouched by market forces will be invaded by private sector entrepreneurship (Public Broadcasting System 2003).

The theory behind the decade-old Edison Project is that capitalism can work for public schools. However, Daniel Gross, author of *Slate's* "Moneybox" column, details the financial practices of Edison including large payments made to its eleven-person board of directors, its generous executive compensation, and immense "sweetheart loans" made to investors that would be unimaginable in a public-school system (Gross 2002). To date, Edison has not made a profit. Edison, which began as a publicly traded company, eventually took the company private (Henriques 2003). The stock, which made its debut on Wall Street at $18.00 a share in 1999 and closed at $36.75 a share in February 2001, fell to fourteen cents a share in October, 2002. Subsequently Edison's founder formed a private equity firm which bought out Edison Schools at $1.76 a share. The receipt of only $1.76 per share was certainly a significant loss for investors who bought into Edison Schools at $36.75 per share (Henriques 2003).

After losing many contracts Edison has diversified away from the management of schools and into marketing supplemental services such as testing, summer school operations and tutoring.

Home Schooling

Although some isolated rural families have taught their children at home for decades, schooling at home took off as a national trend in the mid-1980s (Delbridge 2006). Many parents are not happy with the education taking place in public schools. Homeschooling allows parents to customize education for each child in the family. Each child may be permitted to learn at his or her own pace. A gifted child can be permitted to expand his or her interest in engineering and mathematics. Many of the parents who homeschool find that their children can complete a standard eight hour school day in two hours, leaving time for many extra curricular activities, such as participating in community projects (2006).

Parents of homeschooled children can choose from an immense variety of national curricula, most easily available through the Internet. The Internet itself is also used as an educational tool, as are resources in the community at large. Parents are also aided by support groups. Children who are homeschooled are evaluated annually, using standardized tests given to students in public schools (2006).

Homeschooling is a controversial subject. Many parents choose it in order to place family values, religion, or character building at the forefront of their children's education. It is far easier to draw out the spirituality of one's child in an individual teaching environment than it is in a class full of students. Opponents argue that the purpose of school is to expand children's social knowledge by exposing them to influences from all walks of life and to people and ideas from other religions and cultures. They argue further that children may need to be isolated from their families at times in order to prepare them for life in the outside world. Whatever its pluses or minuses, home schooling is a nationwide trend that continues to grow.

Economics and Poverty in the Public Schools

Moral achievement, in addition to intellectual discipline, is integral to education. Righteousness and virtue go hand in hand with true education (Krishnananda, 2006). Therefore it is incumbent upon us to ensure that each child in the United States has full opportunity both to develop intellectually and to grow spiritually. Children do not have equal opportunity today mainly because public schools in the United States are not funded equally. The federal government provides only eight cents of the education dollar (Hess 2006). Most school funding is provided at the state and local level. The wealth or lack of wealth of the individual school district has a direct impact upon quality of education received by its students. Funding problems are further complicated by the political nature of the issues involved.

In his article, "American Education, Savage Inequalities," Jonathan Kozol details the inequality in funding among New York schools in 1987. Kozol ties poverty to racial isolation, pointing out that the grossly underfunded inner city schools of New York City serve mostly black and Hispanic students (Charon 2005).

The disparity in wealth between black and white Americans is wide. Jan Skutch, a reporter for the *Savannah Morning News* has shared findings from the National Urban League. The net-worth of the average African American family is only *one-tenth* that of the average white family, due largely to differences in home ownership and income. Savannah Mayor Otis Johnson states that the national report reflects the black experience locally. Mayor Johnson believes that education is the foundation of economic security (Skutch 2006).

It is difficult, however, to utilize education as the road to economic security when the playing field among students is not level. In his recent book, *The Shame of the Nation, the Restoration of Apartheid Schooling in America,* Jonathan Kozol presents the results of an exhaustive study of his visits to nearly sixty public schools across the nation (Kozol, 2005). Referencing his 1987 study, Kozol reports that by 2005 the average per-pupil spending in the New York City schools had risen from the 1987 level of $5500 per pupil to $11,700 per pupil. However, by 2005 the wealthier suburban districts had increased their previous per-pupil spending of $11,000 to $15,000 per pupil to as much as $22,000 per year for each student. Moreover the updated dollars had not been adjusted for inflation. When the dollars are adjusted for inflation, the New York City Schools fall even further behind the wealthier suburban districts. Thus, even though the actual dollars spent per pupil have been increased, the disparity among pupils remains. Kozol further reports that the desegregation of black students, which increased continuously from the late 1950s to the late 1980s, has now receded to levels not seen in three decades, noting that the 1972 enactment of a federal program to provide financial aid to districts undertaking efforts at desegregation was repealed by the Reagan administration in 1981. Kozol further points out that the Supreme Court began actively dismantling existing integration programs in 1991.

Jonathan Kozol acknowledges that children living in poor school districts are often victims of other forces which operate against optimum school performance, such as teenage pregnancy, drug use, and single-parent families. He nevertheless points out that if the wealth of New York State were redistributed equally, the education of the children in the poorer districts would be vastly improved. He states that the funding inequalities are the result of a circular phenomenon. In richer districts property lots and houses are more highly valued. Therefore the revenue derived from taxing land and homes is higher. In turn, the reputation of the schools adds to the value of the homes, expanding the tax bases for the public schools, resulting in smaller classes and higher teacher salaries (Kozol 2006).

Kozol's concerns are echoed in great detail by the South Carolina Education Reform Council (2006). In their February 22, 2006, *Report to the Governor*, the Council stated that generating $1000 in local school funds for each school child in South Carolina's poorest district requires a tax of $642 on a $100,000 home and a tax of $1685 on a $100,000 industrial property. Generating $1000 in local school funds in the wealthiest district requires a tax of $78 on a $100,000 home and a tax of $206 on a $100,000 industrial property. The greater burden on the poorer district's taxpayers makes those districts less attractive to the new business investment and economic activity that might help close the gap. Conversely, wealthier districts are better able to utilize state-sponsored economic development incentives, that are funded by all taxpayers in all districts for new or expanding business enterprises, thereby maintaining or even increasing the gap. To partially remedy this concern the Council recommended that the tax structure in South Carolina be reformed (South Carolina Education Reform Council 2006).

The Council's report was published at a time when many South Carolinians were clamoring for property tax relief. It is interesting that in the waning days of the 2005-2006 legislative sessions the South Carolina legislature passed a measure which removed the cost of operating schools from the property tax bill, shifting the cost of schools to the state. Essentially the cost of operating schools became the state's full responsibility. The state is paying the school costs by increasing the state sales tax from five cents on the dollar to six cents on the dollar. This use of the sales tax has been criticized because the poor spend more on necessary items than the rich (Smith 2006a). Therefore the sales tax legislation has obvious drawbacks. It can only be hoped that a more equitable solution will eventually be found.

Another difficult aspect of school funding is the common practice of relying upon local school districts to pass operating levies to run schools and bond issues for construction of schools. Voters, who have little direct say about taxes at the state and federal level, are frequently loath to impose additional taxes upon themselves. The Ohio Education Association reports that only fifty-seven percent of Ohio school issues passed in November 2005 and that less than fifty-one percent in November 2004. The Association reports further that the odds are getting worse for passage of new school levies noting that new operating levies are less likely to pass; school districts are increasing the size of the levies; new operating levies that finally passed in November had already been on the ballot an average of 2.6 times; and many that did pass were decided by razor thin margins (Ohio Education Association 2005).

The reliance upon funding schools through property taxes and local efforts and the complexity of enacting meaningful tax reform at the state level has led to many educators' and school boards' filing lawsuits against their respective states in an effort to achieve school funding equity. New Jersey was the first state to reform its system of funding education via the case of *Abbott v. Burke* (1994). *Abbott* was premised upon an interpretation of the New Jersey state constitution requiring all school districts to be equally funded. Ohio has litigated the constitutionality of its school funding processes for many years (*DeRolph v State* 1997, 1999, 2000, 2001, and 2002; and *State v. Lewis* 2003). The Ohio Supreme Court has found the state funding system to be unconstitutional four times. The legislature has changed the funding structure at least twice during the course of the litigation yet the Ohio Supreme Court has recently found that the funding system remains unconstitutional. Nevertheless the Court has relinquished jurisdiction of the matter, preferring to let the political process run its course (*State v. Lewis* 2003). Massachusetts, South Carolina, Kansas and Vermont have also engaged in extensive litigation over whether or not their states are meeting their constitutional obligation to the individual children of those states (*Hancock* 2005; *McDuffy* 1993; *Abbeville* 1999 and 2005; *USD 229* 1994; *Montoy* 2003 and 2005; *Brigham* 1997 and 2005). The serious student of tax reform can learn much from an examination of these cases.

An effort was made in most of the above cited lawsuits to quantify the exact constitutional standard the involved state is required to meet. South Carolina did not aim high in this regard. The South Carolina Supreme Court has determined that the state constitution requires only that the state prove a "minimally adequate" education to each child (*Abbeville County School District v. State* 1999). When *Abbeville* was remanded to the trial court for clarification of that standard, "minimally adequate" was interpreted to mean the "least possible quantity of a thing that is suitable for the occasion." Trial judge Thomas W. Cooper went on to determine that South Carolina was meeting the "minimum adequacy" standard in all areas except early childhood education (*Abbeville County School District v. State* 2005).

John S. Rainey, executive producer of the documentary *Corridor of Shame* and chairman of South Carolina Board of Economic Advisors, opined that Judge Cooper might have found that South Carolina was not meeting its constitutional obligation in many more areas if the "minimally adequate standard" had not already been set too low (Rainey 2006). Rainey complained that South Carolina already achieves "minimum adequacy" in too many categories, including teen pregnancy, childhood illness and mortality, incarceration rates, one-parent families, drug abuse, obesity, and criminal domestic violence. He further advised that the disease that spawns the former pathologies is poverty, noting that Judge Cooper had, in his decision, pinpointed poverty, not simply education as the problem. Judge Cooper had stated:

> [T]he child born to poverty whose cognitive abilities have been largely formed by the age of six in a setting largely devoid of the printed word, the life blood of literacy and other stabilizing influences necessary for normal development, is already behind before he or she receives the first work of instruction in a formal educational setting. (*Abbeville County School District v. State* 2005, 160-161)

John Rainey sums up the difficulties faced in education in South Carolina by stating "It's the poverty, stupid" (2006). Subsequently the South Carolina legislature has budgeted twenty-three million dollars earmarked for full-day kindergarten for four-year olds from needy families in poor communities. However the amount budgeted has been criticized as being far too low (Brack 2006).

While courts have struggled over the legality of school funding systems, other creative ideas have been espoused to further fund public schools. These include paying the operational expenses of schools from the proceeds of impact fees imposed upon developers in emergent areas and the passage of tax voucher acts to fund scholarships to special tax-supported schools for children from poorer school districts. It has further been suggested that impact fees also be used for the actual construction of new schools (Smith 2006b).

It is likely that as schools struggle to comply with increasingly rigorous state and federal academic performance standards, test results will demonstrate that students in poor school districts obtain the worst test scores. If righteousness and virtue go hand in hand with true education, we need to heed the cries for equality of funding for public schools so that we can draw out the best of the latent faculties in each child.

Ending the Dropout Crisis

In addition to struggling with rigorous academic performance standards schools are struggling to retain more high school students until their graduation. For decades students leaving school prior to graduation have been called "dropouts." There has been much pressure upon schools to decrease the dropout rate.

The first step in lowering the dropout rate is to identify what the rate is (Gates Foundation 2007). Currently it is difficult to identify the percentage of students in the nation who leave high school prior to graduation. Until recently many school districts did not have a standard system to calculate graduation rates. Lack of a standard system leads to wide variations among the states and districts and results in an inaccurate picture of the problem. In reporting their graduation rates some districts count only the number of students who graduate "on time." Under such a method of counting only the number of students who graduate in four years would be reported. The student who takes an additional year in high school or who drops out and later obtains a graduate equivalency diploma would not be included in the "on time" graduation count. Some schools lose track of the students who began high school in the ninth grade and then move away. Such students may or may not graduate somewhere else. Recently the National Governors Association agreed to a common formula for measuring the dropout rate and for identifying dropouts (2007). An accurate measure of the dropout rate should provide a nationwide picture of the problem. Identifying the dropouts should give valuable clues to the solution of the problem. We will know who drops out and why.

Another step toward lowering the dropout rate is to provide students with multiple pathways through high school. Every student will not attend college. Nevertheless each student must be prepared to seek a constructive future. The Gates Foundation has been active in creating such programs. Additional steps to improve the graduation rate include: the increased funding for education which has been provided in the last two years under the *No Child Left Behind Act*; a national public awareness campaign regarding the dropout crisis; and the formation of organizations specifically designed to strengthen the schools (Gates Foundation 2007).

We can hope that increased attention upon the dropout rate and provision of alternate paths through high school will help us end the dropout crisis and thus enable us to better draw out the best and noblest latent faculties in each child.

Religion and the Public Schools

The Pledge of Allegiance

The original Pledge of Allegiance was written in 1892 by Francis Bellamy for a children's magazine (Longley, 2006). Congress officially recognized a revised form of it in 1942. In 1954 the U.S. Congress further modified it by inserting the words "under God." That legislation was supported by President Dwight D. Eisenhower who feared an atomic war between the United States and the Soviet Union. When he signed the bill Eisenhower stated that by inserting the words "under God" into the Pledge, the United States was reaffirming the transcendence of faith in America's heritage and future and strengthening the spiritual weapons which will forever be our country's most powerful resource in peace and war (Longley 2006). Most states have laws regarding requirements for student recitation of the Pledge in schools. Some states require that it be recited, but give students the right to opt out of the exercise. There has been a great deal of debate on the issue of whether peer pressure to say the Pledge constitutes religious coercion, even in states where students have the right to opt out of the exercise.

Michael A. Newdow, a private citizen, filed a lawsuit in 2001 complaining that use of the words "under God" in the Pledge of Allegiance is unconstitutional. The Ninth Circuit agreed, stating that the Pledge is an impermissible government endorsement of religion because it sends a message to unbelievers that they are outsiders, not full members of the political community, and an accompanying message to adherents that they are insiders, favored members of the political community (*Newdow v. U.S.* 2002). Several appeals ensued. Ultimately, in 2004 the United States Supreme Court declined to rule on his challenge to the Pledge, holding that Newdow did not have standing to sue because he did not have custody of his daughter (*Elk Grove Unified School District v. Newdow* 2004).

In conjunction with other parents, Newdow filed another lawsuit in January, 2005. The latter suit resulted in a finding that it was unconstitutional to recite the entire Pledge of Allegiance in public schools because of its inclusion of "under God" (*Newdow v. Congress* 2005). Then, in August, 2005, the Fourth Circuit Court of Appeals ruled the opposite way, upholding a Virginia law requiring public schools to lead a daily recitation of the Pledge of Allegiance (*Myers v. Loudoun County Public*

Schools 2005). Due to divisions among the circuit courts it is likely that the U.S. Supreme Court may be forced to revisit the issue of the constitutionality of the Pledge of Allegiance in the near future. Meanwhile the U.S. Congress has addressed the issue of the Pledge of Allegiance several times. Most recently, in July, 2006 the House passed the *Pledge Protection Act* of 2005. If this bill or similar legislation were passed, such a bill would amend the United States Code with respect to the jurisdiction of federal courts over the Pledge of Allegiance. If such an act is passed, the act's validity will undoubtedly be litigated in the federal courts.

Those who argue to keep "under God" in the Pledge of Allegiance claim that the voluntary recitation of the Pledge is not a coerced religious act. Those who oppose including it in the Pledge claim that its inclusion endorses the religious view that there is a God. Justice Stevens wrote the decision for the U.S. Supreme Court in *Elk Grove Unified School District v. Newdow*. Justice Stevens stated that the recitation of the Pledge of Allegiance is a patriotic exercise designed to foster national unity and pride in the ideals the flag symbolizes. However the issue of the words "under God" in the Pledge of Allegiance is causing division, rather than unity among Americans.

Creationism

Creationism is the belief that the universe and everything in it was created by a supreme being. In modern times the term creationism has come to be associated with the brand of Christian fundamentalism which disputes the theory of evolution.

Until the late nineteenth century, creationism was taught in all schools in the United States (Cheeseman 2008). With the spread of the theory of evolution, public schools began to teach students that humans evolved from a lower order of animals. In the early twentieth century some states, including Tennessee, passed legislation prohibiting the teaching of evolution. In 1927, John T. Scopes was convicted of teaching evolution in violation of a Tennessee law known as the *Butler Act*. His conviction was overturned because the judge had assessed a fine when only the jury had that authority. However the Tennessee Supreme Court upheld the validity of the *Butler Act* itself, reasoning that to prohibit teaching that man has descended from a lower order of animals did not give preference to any religion, and therefore did not violate the First Amendment (*Scopes v. State*, 1927).

Forty-one years after the Scopes trial the United States Supreme Court ruled that the Arkansas law prohibiting the teaching of evolution violated the First Amendment (*Epperson v. Arkansas* 1968). In reaction to the *Epperson* ruling, Louisiana passed a law requiring that public schools should give "equal time" to "alternative theories" of origin. Subsequently the United States Supreme Court held that the new Louisiana law violated the "Establishment" clause of the U.S. constitution. The court stated:

> The Establishment Clause forbids the enactment of any law 'respecting an establishment of religion.' The Court has applied a three-pronged test to determine whether legislation comports with the Establishment Clause. First, the legislature must have adopted the law with a secular purpose. Second, the statute's principal or primary effect must be one that neither advances nor inhibits religion. Third the statute must not result in an excessive entanglement of government with religion. State action violates the Establishment Clause if it fails to satisfy any of these prongs. (*Edwards v. Aguillard* 1987, 582-583)

In *Aguillard* the Court found that the Louisiana law failed all three prongs of the test. However the Court further stated that while creationism is an inherently religious belief, not every mention of it in a public school is necessarily unconstitutional. The court said that creationism may be discussed in school where the intent is to educate students about the diverse range of human political and religious beliefs. The line is crossed only when creationism is taught as science, just as it would be if a teacher were to advocate a particular religious belief (*Edwards v. Aguillard* 1987).

The "Intelligent Design" (ID) movement which started in the 1980s, is an effort to lend scientific validity to creationism. Intelligent Design proponents believe that there is evidence that life was created by an "intelligent designer" because the dynamics of nature and beings are so complex that they must have been "designed" by a superior being, namely God (Y-Origins Connection 2006). Opponents of Intelligent Design state that ID is a pseudoscience because its claims cannot be experimentally tested.

The court battle against the theory of evolution continues. In 2002, six parents in Cobb County, Georgia sued to have the following sticker removed from public school textbooks:

> This textbook contains material on evolution. Evolution is a theory, not a fact, regarding the origin of living things. This material should

> be approached with an open mind, studied carefully, and critically considered. (*Selman v. Cobb County Georgia* 2005, 7)

In *Selman*, U.S. District Judge Clarence Cooper ruled that use of the stickers violated the Establishment Clause of the First Amendment and was therefore unconstitutional. According to Judge Cooper, the sticker sent a message that the Board agreed with the beliefs of Christian fundamentalists and creationists. On appeal *Selman* was remanded to the trial court for additional proceedings. (*Selman v. Cobb County Georgia* 2006). In December of that year, the case was settled when Cobb County school officials agreed not to order the placement of any stickers, labels, stamps, insertions or other warnings or disclaimers containing language similar to the language that was placed on the stickers that led to the lawsuit. School officials also agreed that they would not take other actions that would undermine the teaching of evolution in biology classes (National Center for Science Education 2006).

In December, 2005, federal judge John E. Jones III ruled that the school board of Dover, Pennsylvania, violated the Constitution when it sought to teach Intelligent Design. Judge Jones stated, "In making this determination, we have addressed the seminal question of whether ID is science. We have concluded that it is not, and moreover that ID cannot uncouple itself from its creationist, and thus religious, antecedents" (*Kitzmiller v. Dover Area School District* 2005, 136).

The battle continues in churches, legislatures and school boards. Religious activists urge parents to remove their children from the public schools in order to avoid having their children exposed to Darwinism. In May, 2006, a South Carolina House committee considered an amendment which would require school textbooks, software and other instructional materials to be selected based upon whether the materials "critically analyze" the theory of evolution (Drake 2006). Drake reports that biologists, claiming that an understanding of evolution is crucial to understanding how organisms became what they are today, oppose "critical analysis" as a backdoor way to insert intelligent design into the curriculum (2006). Meanwhile in Ohio, the State Board of Education voted in February, 2006, to get rid of curriculum guidelines that would have allowed the teaching of intelligent design in Ohio schools (Y-Origins Connection 2006, 2).

Intelligent Design is therefore the new "creationism." Academics believe it is a sham. Fundamentalists see it as an issue of morality. In the near future, school districts will continue to struggle over whether they must reject scientific findings that are at odds with Biblical teachings.

Future Trends—2020

Former Superintendent of Beaufort County, South Carolina, School District Edna H. Crews was consulted by this writer on the issue of what public schools may be like in 2020. Crews has more than thirty years' experience as a teacher, educational counselor and administrator. Her response was that the schools of 2020 must prepare students to work for at least fifty years—certainly past age 65—because as the population lives longer, the economy may not support early retirement. In 2020 we may also need to keep students in school longer in order to provide the additional learning they will need in a more complex society and in order to accommodate the expanding base of knowledge. The curriculum's math content alone doubles every four years. Crews states that most students will have a number of careers in their lives. Their education must provide them with flexibility and skills to negotiate such career changes. Students of 2020 will differ from students of today in that they will initially enter school with a wider base of experiences. Many of the parents of children attending school today are a lot older than the parents of students in previous generations and so may be financially better off. Unlike yesterday's children many entering school today have already traveled extensively and have flown in airplanes. Undoubtedly children entering school in 2020 will have had even broader experiences (Crews 2008).

Perhaps the most crucial thing we can teach students of current and future generations is how to distinguish fact from opinion. In 2020, students will be bombarded with information from all directions. They must be able to evaluate the information they encounter in order to make an accurate assessment of which information to trust. More than ever before, the teaching of critical thinking skills will be a priority.

It is likely that the schools of the future will employ "master teachers" whose mastery of subject matter will far exceed that knowledge of the average classroom teacher. Crews reports that the State Department of Education of South Carolina already has a model program for the education of master teachers.

Most teachers of 2020 will likely have the opportunity to earn "merit pay" for excellent performance. Beaufort County already has three pilot schools in which the practice of merit pay has been instituted. There may also be a severe teacher shortage in 2020 because the colleges are not graduating enough future teachers. Teaching is often a second career for younger retirees, particularly in such places as Beaufort, South Carolina, where several military bases are located. All states need alternatives to traditional educational certification for teaching. Crews points out that the State of South Carolina has a program allowing an individual with a bachelor's degree to be employed as a teacher without having taken any courses in education. Once employed, the new "teachers" have three years to obtain the necessary courses in which to obtain certification in their field. Meanwhile, they receive support in classroom management from the school district where they are employed (Crews 2008).

On the issue of technology, Crews believes that by 2020 there will be many more "virtual schools" as well as a need for computers in the home and on the desktop of every student. In 2020 there will be greater use of "Advanced Placement" courses and "Master" teachers to provide televised courses throughout the school district and into the community at large.

The schools of the future will most likely focus on individualized instruction and move away from grade levels. Crews states that a child who fails any grade three times is eighty-five percent less likely to graduate from high school. Beaufort County has three schools which offer an "International Baccalaureate" (IB) diploma. The curriculum of an International Baccalaureate school can yield a diploma which gives international recognition to the curriculum at the school. There may be many more IB schools in the future as we proceed further toward a global economy. Our educational system was designed for an agricultural society that needed children to be available for farm work during the summer. A school system retooled for today's world must focus more upon opportunity than simply upon graduation from high school.

In summary, if righteousness and virtue go hand in hand with true education, it is certainly incumbent upon us to refine our systems of funding education delivery so that every student has a real opportunity to achieve. Educational reform has been caught up in the political process for more than fifty years. Tax vouchers, charter schools, the *No Child Left Behind Act*, and privatization of education have not solved the problems. If we believe in equality and believe that education is the responsibility of the state, it is time for a new type of political process, a process rooted, not in partisan politics, but in the hearts and minds of those that believe in the "American dream" and reason that the achievement of this dream should be within the grasp of every citizen.

By 2020 the growth of a global economy and the national interest may require a greater federal role in education and perhaps a national curriculum, such as the International Baccalaureate curriculum. Pre-kindergarten education will likely become mandatory. Parents will have more choice in which schools their children actually attend. At the same time, the public schools themselves will offer more educational choices to students. The growth of charter schools and the home schooling phenomenon has made it clear that many parents do want religious and ethical instruction in the lives of their children. There will be increased accountability for funds spent and results achieved. New school buildings will be constructed so make best use of new technology. Virtual instruction will be provided in most schools and cyber schools will be important in remote areas.

In our zeal to be technologically proficient, it may be helpful to remember that the very technology we prize can also be used to our detriment. Whether the purpose of education is to "draw out the best, sublimest and noblest of the latent faculties in man," or whether it is to "serve the ruler" by producing responsible citizens who will strengthen the nation, instruction cannot be all academic. Our world is filled with wars, refugees, disease and violence. Now, as never before, we are in danger of destroying our world. If there is no place in our educational system for instruction in human values and conflict resolution, all of our teaching may be for naught.

References

Billingsley, Peter. (2006). "The Virtual Classroom." Asheville, NC: Mountain Xpress. www.mountainx.com. (accessed November 25, 2006).

Brack, Andy. (2006). "Early Education a Good Investment for South Carolina." *The Island Packet*. Bluffton, SC. April 24.

Cheeseman, Robert H. (formerly with the Mississippi Sate Department of Education). (2008). Personal Interview with the author. 114 Doncaster Lane, Bluffton, SC.

Crews, Edna H. (2008). (consultant and former Superintendent of Beaufort County School District). Personal interview with the author. P. O. Box 1021, Beaufort, SC.

Cohen, Hal. (2005). "The Future is Now." www.midatlantic.construction.com/features/archive/0510_Feature2.asp. (accessed May 5, 2006).

Collins, Gale. (2006). "Reining in Charter Schools." New York: *New York Times*. May 10. www.nytimes.com. (accessed November 25, 2006).

Delbridge, Rena. (2006). "When School's At Home." *Casper Star-Tribune*. Jackson Hole, Wyoming. May 13. www.jacksonholestartribune.com. (accessed May 17, 2006).

Drake, John C. (2006). "Debate in Teaching Evolution Resurfaces." *The Christian Post*. Washington, DC. May 16. www.christianpost.com/article20060516/7910.htm. (accessed November 30, 2006).

Feller, Ben and Andrew Welsh-Huggins. (2006). "No Child Law Lining Some Pockets." *The Island Packet*. Bluffton, SC. April 4.

Fulghum, Robert. (1988). *All I Really Need to Know I Learned in Kindergarten: Uncommon Thoughts on Common Things*. New York: Ivy Books.

Edison Schools. (2006). www.edisonschools.com. (accessed May 16, 2006).

Gates Foundation. (2007). "Summit Propels National Movement to End America's Silent Dropout Epidemic." www.gatesfoundation.org/UnitedStates/Education/TransformingHighSchools/Announcements/Ann. (accessed September 21, 2008).

Gross, Daniel (2002). "Edison's Dim Bulbs." *Slate*, August 30, 2002. www.slate.com. (accessed May 19, 2006).

Henriques, Diana B. (2003). "Edison's Schools' Founder to Take It Private." New York: *New York Times,* July 15. www.nytimes.com. (accessed May 19, 2006).

Hess, Frederick M. (2006). *Tough Love for Schools*. Washington, DC: American Enterprise Institute for Public Policy.

Hogan, Kevin. (2006). "Is this the School of the Future?" www.scholastic.com/administrator. (accessed May 5, 2006).

Kozol, Jonathan. (1991). "American Education: Savage Inequalities." In *Social Problems: Readings with Four Questions* 2nd Ed., eds. Joel M. Charon and Lee Garth Vigilant. Wadsworth/Thomas. Belmont, CA. 2006

Kozol, Jonathan. (2005). *The Shame of the Nation, the Restoration of Apartheid Schooling in America.* New York: Crown Publishers.

Krishnananda, Swami. (nd) "The Spirit of Education." www.swami-krishnananda.org. (accessed August 9, 2006).

Lee, Carmen J. (2000). "Education 2000: Reforming Schools for a New Century." www.post-gazette.com/regionstate/20000903future3.asp. (accessed January 9, 2006).

Lips, Dan. (2005). "Grading No Child Left Behind." www.fox news.com. (accessed November 21, 2006).

Longley, Robert. (2006). "Brief history of the Pledge of Allegiance." www.usgovinfo.about.com/csusconstitution/a/pledgehist.htm. (accessed November 21, 2006).

Microsoft (2006). "School of the Future: Understand the vision." www.microsoft.com/Education/SchoolsofFutureVision.mspx?pf=true. (accessed May 7, 2006).

Microsoft. (2007). "Bill Gates, Chairman, Microsoft Corp." www.microsoft.com/presspass/exec/billg/bio.mspx?pf=true. (accessed September 24, 2008).

National Center for Science Education. (2006). "Americans United Applauds Settlement of Georgia Lawsuit over Evolution Disclaimer." www.ncseweb.org/resources/news/2006/GA/272_selman_v_cobb_county_settled_1. (accessed September 24, 2008).

National Commission on Excellence in Education. (1983). *A Nation at Risk.* Washington, DC: U.S. Government Printing Office.

Ohio Education Association Research Bulletin. (2006). "OEA Review of Cost of NCLB Implementation Study." www.ohea.org/researchReports/schoolFinance.aspx. (accessed May 20, 2006).

Ohio Education Association *OhioSchoolsOnline*. (2005). "School Funding stuck on a Treadmill." www.ohea.org/newIssues/OhioSchools.aspx. (accessed May 20, 2006).

Public Broadcasting System. (2003). "Frontline Public Schools Inc.: Inside Edison's Schools." www.pbs.org/wgbh/pages/frontline/shows/Edison/inside/. Webcast July 3, 2003. (accessed May 17, 2006).

Rainey, John S. (2006). "['Minimally adequate'] a curse for state and education system." *The Island Packet*. Bluffton SC. January 15, 2006.

Saunders, Doug. (2002). "For-Profit U.S. Schools Sell Off Their Textbooks." Common Dreams News Center. www.commondreams.org. (accessed May 17, 2006).

Skutch, Jan. (2006). "Black Americans have only a 10th of what white Americans do." *Savannah Morning News*, Savannah, Georgia. April 4, 2006.

Smith, Janet. (2006a). "Property tax relief plan misses the bigger picture." *The Island Packet*. Bluffton, SC. June 5, 2006.

Smith, Janet. (2006b). "Numbers get more daunting as costs of growth get clearer." *The Island Packet*. Bluffton, SC. January 29, 2006.

South Carolina Education Reform Council. (2006). "Report to the Governor of the South Carolina Education Reform Council." Budget and Control Board of the State of South Carolina, Columbia, SC.

Wallace, Kathryn. (2005). "America's Brain Drain Crisis." *Reader's Digest*: Pleasantville N.Y. www.rd.com/content/opencontent.do?contentId=24076. (accessed November 30, 2006).

Y-Origins Connection. (2006). "Questions and Answers about Intelligent Design." www.y-origins.com. (accessed August 31, 2006).

Cases

Abbeville County School District v. State, 335 S. C. 58, 515 S. E. 2d 535 (1999).

Abbeville County School District v. State, (2005). www.scschoolcase.com/Abbeville-County-Order.pdf. (accessed November 24, 2006).

Abington v. Schempp, 374 U.S. 203 (1967).

Abbott v. Burke, 136 N. J. 444 (1994).

Brigham v. State, 692 A. 2d 384 (1997).

Brigham v. State, 889 A. 2d 715 (2005).

Brown v. Board of Education, 347 U.S. 483 (1954).

DeRolph v. State, 78 Ohio St. 3d, 193, 677 N. E. 2d 733 (1997).

DeRolph v. State, 98 Ohio Misc. 2d 1, 712, N. E. 2d 125 (1999).

DeRolph v. State, 89 Ohio St. 3d 1, 728 N. E. 2d 993 (2000).

DeRolph v. State, 93 Ohio St. 3d 309, (2001).

DeRolph v. State, 97 Ohio St. 3d 434, (2002).

Edwards v. Aguillard, 482 U.S. 578 (1987).

Elk Grove Unified School District v. Newdow, 542 U.S. 1 (2004).

Hancock v. Commissioner of Education, 443 Mass.428 (2005).

Kitzmiller v. Dover Area School District, (2005). www.pamd.uscourts.gov/kitzmiller/kitzmiller_342.pdf. (accessed November 21, 2006).

McDuffy v. Secretary of the Executive Office of Education, 415 Mass. 545, 621 (1993).

Montoy v. State, 275 Kan. 145, 62 P. 3d 228 (2003).
Montoy v State, 279 Kan. 817 (2005).
Myers v. Loudoun County Public Schools, 251 F. Supp. 2d 1262 (2005).
Newdow v Congress, 383 F. Supp. 2d 1229 (2005).
Newdow v. U.S., 292 F. 3d. 597, Ninth District, California (2002).
Scopes v. State, 289 S. W. 363, Tenn. (1927).
Selman v. Cobb County Georgia, (2005). www.talkorigins.org/faqs/cobb/selman-v-cobb.html. (accessed November 27, 2006).
Selman v. Cobb County Georgia, (2006). www.call.uscourts.gov/opinions/200510341.pdf. (accessed November 27, 2006).
State v. Lewis, 99 Ohio St. 3d 97 (2003).
U.S.D. No. 229 v. State, 256 Kan. 232, 885 P. 2d 1170 (1994).
Zelman v. Simmons-Harris, 536 U.S. 639 (2002).

Statutes

U.S. Code 20 (2002) §§ 7801 et seq.

Chapter 4

Politics

ROY GODWIN, PHD

> We hold these truths to be self-evident, that all men are created equal, that they are endowed by their Creator with certain unalienable Rights, that among these are Life, Liberty, and the pursuit of Happiness. . . .
>
> (In Congress, July 4, 1776, The Unanimous Declaration of the Thirteen United States of America)

The Social and Spiritual Aspects of Politics

In the early years of the twenty-first century the United States finds itself aging, not so much in the sense of the population's moving into older adult years, although that is true. It has more to do with the time that we as a nation have been on the scene of passing history. Some may characterize the United States in comparison with other nations in the world as being still in our adolescence in terms of how long we have been around and in our behavior. We are fast to react with belligerence when we are attacked without spending necessary time to ask questions or engage in much self-reflection. Our war on terrorism and the current war in Iraq to establish democracy there may serve as an example.

At this time it would serve us well to try to get more understanding of our circumstances politically in the light of current trends. Let me first put forth a disclaimer: I am not a politician, nor the son of a politician. I am a member of the ordained clergy—one who served in the ranks of an evangelical denomination, the Southern Baptist Convention, for over forty years as pastor, professor, and longest, as a denominational administrator. During that time I served mostly in large metropolitan areas, anywhere from the inner city to the rural urban fringe. Almost twenty of those years were in the metropolitan Washington, DC, area. My experience with government—from national to municipal—has varied from advocate to adversary, depending on the action of the political machine. At least some of what I have learned in that fray is reflected in what follows in this chapter.

So that the reader may understand my perspective from experience as well as my formal training in theological ethics and the sociology of religion, some definitions of terms would be helpful. *Politics* has to do with the art and science of how we govern ourselves, the total complex of relations among people in society. It includes formal structures of the three branches of national government and those of the state, county, and municipal governments, both elected and appointed, and their actions, practices, and policies. Political parties are also included as well as formal and informal special interest groups and volunteer organizations among us that may participate in the decision-making process.

Social and *spiritual* aspects refer to those ideas, forces, movements, and trends, from the smallest to the largest, that are present in exchanges among people within the community. The primary focus is not only upon social trends, but especially the spiritual dimension of such trends—the *soul* of politics. That dimension may include, but not be limited to, religious trends. It is an expression of that which is from the highest dimension of ultimate reality to the depths of one's being. As we examine our subject we will use gleanings from several disciplines, including sociology, political science, theology, and history.

Ethics is that which we define as the highest good. It is the effort of free persons to understand the meaning of the good as it is conceived in the harmony of parts of the whole. It is also the ethical action in the effort of free persons to achieve the good which is conceived as the harmony of self-interest with the larger social whole. Autonomy, freedom of the will, is essential in understanding ethical behavior which is preferably from the free choice of free persons in a free society (Johnston 1981).

Megatrend: The Rising Tide of Spirituality

The major trend of growing interest in spirituality is the proverbial tide that raises all ships. It is evident in society in general and in business and politics in particular. While modern science with its "culture of separation" has produced fragmentation within its own ranks of scientific exploration (Bella, Madsen, Sullivan, Swidler, Tipton, 1996), there is emerging, especially from the ranks of quantum physics and neuroscience, a movement toward wholeness (Zohar and Marshall, 2000). Such wholeness carries with it a flavor of spirituality not at odds with science but rather quite compatible, an "at-one-ness," which leads to the major trend at the beginning of the twenty-first century: the rise of spirituality. Such spirituality is pervasive, moving into the corporate world of business, economics, education, politics, and, of course, religion.

Furthermore, spirituality is not simply an intuition or dream that can be dismissed as something ethereal. It is a real mystical experience that can be verified as biologically observable and scientifically real. At its roots it is intimately interwoven with human biology which in some way compels the spiritual urge. From their studies of the functions of the human brain during spiritual mystical experiences, Andrew Newberg and Eugene d'Aquili conclude that a neurological process has evolved which allows humans to transcend material existence and knowledge and to connect with a deeper, more spiritual part of themselves, one that is perceived as an absolute, universal reality connecting all that is. What results is a kind of "interspirituality" in which, through dialogue and reconciliation, all religions may travel along different paths toward the same goal of wholeness and unity. Such interspirituality has profound pragmatic implications as well, since mysticism allows us to transcend egotistical fears of alienation from one another and provide a sure foundation for a better world (Newberg and d'Aquili, 2001). Such technological and biological advances suggest that our psychological and physiological capacities are far greater and much more sophisticated than we ever imagined. This leads futurists Joel Barker and Scott Erickson to conclude that God, Mother Nature, or evolution, has endowed us with extraordinary capabilities to attend to our real needs—not material needs, but needs of the spirit, our internal states: our hopes, fears, dreams, relationships, and goals (2005).

By spirituality we mean something more than religion although the term may include religion. It is expressed by some who say, "I am spiritual but not religious," usually meaning that they have great respect for and may even worship an infinite Being who may be the Creator or "ultimate ground of being," and they may have a sense of reverence of that Presence immanently within as well as transcendentally. They may not have membership in a local church or attend worship within a congregation with any regularity; however, some may gather in small groups with others of like mind. Ken Wilber argues this to be the case when one may simply be speaking for one's self in describing what one believes personally. As soon as one formalizes a routine, gathers with others of like mind and interest, or tries to pass on from one generation to another one's beliefs, one moves into the category of "religious" (2005). Others might express hostility toward organized religion and prefer to keep their distance. Wade Clark Roof describes spiritual and religious attitudes within the Baby Boomer Generation, the leading generation in this modern movement, as "seeking,"—a quest usually followed by reflection, and then, by some action in an outward direction toward social values. In fact, Roof reminds us, it was Boomer Vice President Dan Quayle who introduced the concept "family values" into our public debate (1999).

Paul Tillich defines Spirit most profoundly as "first of all power, the power that drives the human spirit above itself toward what it cannot attain by itself, the love that is greater than all other gifts, the truth in which the depth of being opens itself to us, the holy that is the manifestation of the presence of the ultimate" (1963, 84).

To this let us add Patricia Aburdene's description of Spirit as the attribute of the Divine which dwells rather privately within humanity, analogous to the Holy Spirit but in a more interfaith, ecumenical, and nondenominational way. It is distinguished from religion, which is more behavioral and experiential in character, and is usually formal and publicly organized. She concludes that the "Power of Spirituality," the quest for spirituality in its desire to connect with God, the Divine, the Transcendent, is the greatest megatrend of our era (2005).

Now let us move another step to the role of spirit in society. All natural and living systems can exist in one of four states: closed, steady-state, complex adaptive, and turbulence. No human system can exist without three kinds of capital: material (raw materials), social (relationships, community customs, and institutions), and spiritual (meaning, values, purpose). The closed societal system is one locked into life and customs largely cut off from the ebb and flow of the larger society, usually agrarian much like the Amish communities. The steady-state system is one in which there is some movement and some inward and outward flow that usually is in steady balance. Some participants travel outside the system, and some tourists come in. Some material needs are imported, and some products are exported. The population draws on the spiritual capital of perhaps several belief systems which are usually long-held, nonproselytizing, unchanging, and live side by side. Spiritual capital is invested in strong social and cultural traditions that guide the lives of the vast majority of the people while allowing a few artists and writers to be creative within the familiar boundaries of tradition. In both the closed and the steady-state systems the spiritual capital is static and passive as it is passed down from elders. Usually it is to traditional family values that members are called to return, the ideals of a steady-state system in balance. Turbulence is a state about which little can be done because things have disintegrated and are in a state of anarchy.

As an open capitalist democratic society that is sustainable and able to maintain itself and evolve into the future, we recognize that ours is a special kind of entity that scientists call a complex adaptive system, or a self-organizing complex system. That is to say that matter and energy and information flow freely throughout, and are organized from the inside as we are poised at the edge of chaos, a critical point beyond equilibrium. An input of deeply transpersonal values and moral principles of conscience from within is necessary for our human sustainable system's self-organizing capacities.

As we progress in these dynamics we move from one state to another, a shift that scientists call a phase change such as that transition from a traditional society to a modern society. An example would be the rise of capitalism in the eighteenth century coupled with the Industrial Revolution resulting in the complex adaptive system that persisted throughout the twentieth century. Following the turbulence and instability of two world wars, we have experienced the sapping of Enlightenment ideals and the slow disappearance of Christianity and its values and relationships from most people's lives, except perhaps in Third World countries where Christianity is on the rise. That has been followed by greed, materialism, and selfishness, the information revolution, globalization, and the diminution of the nation state. Danah Zohar and Ian Marshall conclude that we now stand on the brink of a profound phase change in which we can spin off into disintegrating turbulence, or we can become a new complex adaptive system. If we choose the latter, we must use our spiritual intelligence with its essentially persisting but constantly evolving structure—its relationships, customs, mores, institutions, and the like, that can constantly rearticulate abiding values and purposes, or create new values and purposes that can guide us in a new life of recreated form (2004). Perhaps the people group that will lead us through this phase change is the group identified as the "Cultural Creatives" as described by Paul Ray and Sherry Ruth Anderson. They are the over twenty-six percent of the U. S. population and from thirty to thirty-four percent of Western Europe who value nature, authenticity, spirituality, peace, relationships, feminism, social justice, and social responsibility. They are neither traditionalist nor confined to a demographic slice of the population such as Baby Boomers. They are held together by their values, and through zealous commitment to them they are creating a new culture (2000).

It is against the backdrop of this major trend of the rise of spirituality specifically in the American society that we view the next major trend. The rise of spirituality sets the stage for discussions related to the soul of politics, moral and ethical behavior in pursuit of the highest good, and the defining of social values.

Megatrend: The Movement Toward Fundamentalist Politics

At the time of this writing, the five hundred pound gorilla in the living room of United States politics is the war on terrorism. It is a profound reality, but it serves also as a symbol whose meaning moves from retaliation toward terrorists to a conflict between democracy and theocracy. The desire to spread democracy through the use of military might has become a tactic aimed to free the whole world from oppression and tyranny.

Scholars and prognosticators were eager to project both serious concerns and optimism as they peered into the twenty-first century from the other side of the terrorist attack on September 11, 2001. Since that time a very serious sense tempers any optimism at large by futurists, except perhaps for fundamentalist dispensational preachers who see the beginning of the end of history as we know it and the subsequent apocalyptic deliverance of the faithful after the Battle of Armageddon on the burnt and bloody sands of the Holy Land. But before we get into that discussion let us explore the bigger picture of what has been happening in American society leading up to that war.

The Soul of Politics

As in religion, so it is in politics. To understand the basics, the soul of politics, one must start with its sacred scriptures. In the case of the United States, those documents are the *Declaration of Independence* and the *Constitution*. They are to be taken together because the former gives vital commentary to the latter. Beyond the study of those documents one should give attention to the historical context in which they were crafted. That study should resist temptations toward *eisegesis*, the reading into the texts one's own ideas or what one wishes had been there in its original meaning.

When a party that has been out of power finally comes into power, it often attempts to keep that power by changing the rules by which it arrived at that new status so as to prevent the opposition from following in their footsteps to unseat them. So laws are changed, new ones are passed, and the constitution is reinterpreted or changed. This latter success changes the foundation into a basis for reform by the current powers that be. That trend is and has been happening in our government for years with a few temporary interruptions from time to time.

One may begin with the debate about the spiritual nature of the Constitution of the United States. Some argue that the founding fathers were devout religious men who were inspired by the Creator to set forth certain articles for constituting the new government and setting forth a list of certain unalienable rights of its citizens. Looking at the historical context, professor Elaine Pagels argues that in 1776 what the *Declaration of Independence* states as self-evident truths, i.e., "that all men are created equal and are endowed by their Creator with certain unalienable Rights . . ." was relatively and radically new. It is the conviction that the individual has intrinsic rights, certain claims on society and even claims against society, that the state must recognize and protect in order for it to be legitimate. Where did this idea come from? Pagels declares that it came from the Genesis account of creation, the Gospel of John, and from George Fox and the Society of Friends who believed in the light of God in every person (2005).

Judge Andrew Napolitano states that in order to understand the Constitution and where it came from, one must understand the Declaration of Independence as well as the Natural Law which grounds both documents (2006). According to Natural Law theory, the law extends from human nature, which is created by God. It is only natural that all human beings desire freedom from artificial restraint and natural that all human beings yearn to be free because our freedoms stem from our very humanity—and ultimately from the Creator of humanity. The Declaration of Independence refers specifically to God-given rights which are received by virtue of our humanity. Such understanding of Natural Law in regard to the conferring of rights upon humanity and the relationship between those rights and the role of government is fundamental to our understanding of the Constitution. It follows then that when the people created state legislatures and they in turn created the Congress, they never gave those bodies the authority to interfere with natural rights. In fact, no one can take away another's natural rights except by jury following due process. Furthermore, Natural Law recognizes that not all rights are natural and that some rights do come from the state.

On the opposite side is the Positivist theory which states that the law is whatever those in power say it is, whether their decision is democratic or dictatorial in nature. It asserts that all laws must be written, and that the government can enact whatever laws it wishes without restraint. There is no higher law. The majority always rules and always gets its way except when it chooses to condescend to the minority. Since rights come from government and can be repealed at any time, the door is open to tyranny.

Napolitano concludes that currently in the United States Positivists are carrying the day. The government violates the law while at the same time it busily passes more legislation to abridge people's liberties. Due to the lack of restraints, the government is fairly free to enact whatever laws it wishes. A feature of our Constitution that was unique at the time of its writing was the system of checks and balances that assured distribution of power among the branches of government. Furthermore, enacting a Bill of Rights strengthened the balance of powers. The power now seems shifted to the executive branch as George W. Bush, and other former presidents, have claimed the privileges of Commander in Chief during wartime. Laws have been recently passed that encroach upon the rights of citizens, for example, the *USA PATRIOT Act*, wire taps, searches without prior warrants, and the suspension of *habeas corpus*. Powers that usually have been retained by states—a system of federalism that preserves the autonomy of the states—have been intruded upon in the spread of federal power, often through the strong-arm tactic of granting or withholding federal spending.

The Supreme Court is often called upon to apply the Constitution to subjects and disputes that the Founders never imagined: abortion, gay rights, flag burning, and prayer in public schools, to name a few. Such controversies usually center on the question: should the Court treat the Constitution as a living document, allowing judges to interpret it according to the times, or should they yield all deference to the rule of the majority in the legislature? (Napolitano 2006).

The modern critical debate began with the administration of President Ronald Reagan and the emergence of conservative power. Edwin Meese, III, President Reagan's first attorney general, in 1985 called on judges to return to what he called "a jurisprudence of original intention" in interpreting the Constitution. An example of how the court took a turn from previous stances is in the matter of separation of church and state. Meese's challenge was taken up by then Associate Justice William H. Rehnquist, who in his dissent in a case from Alabama known as *Wallace v. Jaffree*, argued that Thomas Jefferson's image of a wall of separation between church and state is only a "metaphor based on bad history" which should be "frankly and explicitly abandoned." Since Jefferson was absent from the constitutional convention in 1787 and from the first Congress of the United States when the Bill of Rights was enacted two years later, Rehnquist continued, we must turn to James Madison for guidance on what the framers intended, and Madison did not care if the constitution contained religious guarantees at all or that there be strict separation of church and state. Many feel that Judge Rehnquist was wrong in his interpretation of the history surrounding those events in arguing that the intent was simply to forbid the establishment of a national church and to prevent the preferential treatment of one Christian sect over another. The record does indeed demonstrate decisively that after three attempts to use language that is more akin to that of Rehnquist's interpretation, the framers clearly and decisively intended nothing less than the institutional separation of church and state by adopting the language: "Congress shall make no law respecting an establishment of religion. . . ." However, the issue no doubt will continue to be pressed by the revisionists especially since the recent change in the makeup of the Court (Hastey 2000).

One may conclude at this point in regard to the controversy over the correct interpretation of the Constitution and its application to governing the land, if the soul of politics is not lost, it is exceedingly troubled. There will continue to be battles over its meaning and assaults on the people's rights that it was crafted to protect. In fact there was a movement initiated to take legal action against the Bush administration over what was interpreted by some as grounds for impeachment (Lindorf and Olshansky 2006).

The March to the Right

How did we get to this place in our march toward the political right? It seems that streams of upsurge and decline in both politics and religion have intersected to bring us here. First was the decline of liberalism. Professor Robert Wuthnow reminds us that the liberal strand comes out of long traditions of church-state relations in Scotland, England, and northern Europe in which religious leaders made statements on public issues. They were joined later in America by those from the Catholic regions of Europe. To the influence of the Enlightenment, especially Lockean contractualism and the communal liberalism expressed in Rousseau's ideas of civil religion, one may add the influence of deism and rationalism to form building blocks of liberal public theology. As it moved into the public sphere it embedded itself in the wider notion of human rights and responsibilities applying moral principles on which people of goodwill could agree. From a subjectively knowable reality they moved to an external manifestation of a lawful and beneficent order; they concluded that through reasoned deliberation people of goodwill can be expected to achieve an external manifestation of a lawful and beneficent order and agreement about the common good. So moral order is not something to be imposed externally through coercion, but comes voluntarily through free choice by an informed citizenry that includes people of varied national and religious heritage, ethnicity, race, religion and political persuasion (Wuthnow 1993).

At the beginning of the twentieth century, liberal Protestantism, mostly in mainline denominations, was awash in the Social Gospel movement and optimistic about our country's future as it found a government open to its input in defining the most crucial values for our society. Even though two world wars shattered optimism about the nature of humankind, still there was hope as the Democratic party embraced its values in their political platforms of the New Deal to bring us out of the great depression and later of President L. B. Johnson's Great Society to bring us together in the achievement of civil rights for all people regardless of race, creed, or gender. As they were getting their social agenda enacted, there began a decline in membership and influence of mainline denominations. The 1960s brought much social unrest with the Viet Nam War and social upheaval around the morals of major segments of our society, followed by Supreme Court decisions such as *Roe v. Wade* that divided the nation on the issue of the value of human life. The Democratic party began to slip in its grip on the controls of political destiny, and mainline Protestants were left making mostly prophetic pronouncements.

With the shift of social concern into the government services and the decline of liberal mainline churches there was a corresponding rise in influence of the conservative evangelical churches in the social arena. This branch of Protestantism which had been marginalized throughout much of the early twentieth century began in the 1970s to rise in number and prestige with the emergence of the "neo-evangelicals" (Neuhaus 1981).

Adopting a more pessimistic view of human nature, conservatives also assume that some subjective sense of a beneficent moral order can be found, but they place greater emphasis on the externality of that order and limit its discovery to a more restricted sphere. They believe that the chosen regenerate few claim divine insight mostly through learning principles that have been laid down and by paying heed to the institutions in which these principles are understood. They are more reactionary to their liberal counterparts and were often first formed from splinter groups who separated from the mainline denominations. They oppose secular optimism that is evident in academic settings which they characterize as moral relativism, and against that same influence found in legal and political debates. Believing that individuals do not have a reliable moral sense inherent within themselves, it is therefore necessary to have strong churches, strong moral instruction in the schools, and legal sanctions to ensure public decency.

Recognizing that it is important to have a myth of origin, one may argue that the United States had originally espoused these same ideals, and the early Puritan colonies had a theocratic orientation as they were founded on strong biblical principles. There was a special covenant between this new nation and God, and many of the founders were dedicated believers. From this lofty height of moral insight we have degenerated into having religious and secular institutions that have turned away from the biblical order. Nevertheless the United States still has a mission to fulfill in fighting godless ideologies, such as communism and radical Islam, and keeping our country free from moral decay. Even though the conservatives have been skeptical of government intrusion in religious matters, they support a strong system of legislated morality (Wuthnow 1993).

Following the early measured successes of the Moral Majority in the twentieth century, in the early twenty-first century the evangelicals and fundamentalists have emerged with great force and influence in the political arena. They, like their liberal counterparts in the early twentieth century who aligned with a political party, have now aligned themselves with the Republican Party. They have enjoyed progress along with their politically conservative allies in seeing much of their social program codified and implemented. The political right was claiming dominance in all three branches of government and moving toward the status of dominant national party until the elections of 2006 revealed a pendulum swing that seemed to reverse this trend. The question now is whether or not the evangelical influence will remain or decay as the political pendulum begins to swing back toward the middle as was the case with liberal Protestants earlier.

The attack by the terrorists on September 11, 2001, has presented an opportunity to make major changes in the way the Constitution is administered. It is now caught in the crossfire of the two warring parties of our political system. The major charge has been coming from the far right of the conservative Republican party, which was introduced in the Nixon years but began in earnest with the Reagan administration. It continued in the George H. W. Bush administration, was followed by a pause during the Clinton administration and Republican Congress, and then regained full speed with the George W. Bush administration, a Republican controlled Congress, and a majority of judges on the Supreme Court.

At the risk of oversimplifying, typical conservative beliefs include at least three things: small government with low taxes, traditional values, including the sanctity of life, and a hawkish foreign policy (Barnes 2006). In a more thorough attempt to describe the nature of conservatism, John Dean, former White House counsel to President Richard Nixon, gives a brief history of the conservative movement from classical conservatism to modern conservatism as defined in the thought of former Senator Barry Goldwater. In sum, the essentials include: "draw on the proven wisdom of the past; do not debase the dignity of others; and maximize freedom consistent with necessary safety and order" (Dean 2006, 18). Dean adds that absent from Goldwater's conservatism was any thought of the government's imposing its own morality, or anyone else's, on society.

He continues in subsequent chapters to discuss the development of conservatism with its current link to authoritarianism which plays out with strong authoritarian leaders who display these traits: "dominating, opposed to equality, desirous of personal power, amoral, intimidating, and bullying; some are hedonistic, most are vengeful, pitiless, exploitive, manipulative, dishonest, cheaters, prejudiced, mean-spirited, militant nationalistic, and two-faced" (2006,183). Those authoritarian leaders are able to sustain their leadership "because contemporary conservatism has generated countless millions of authoritarian followers, people who will not question such actions." What emerges is an administration against which there are little or no constraints, determined to dominate the world with America's best interests as they define them, and that uses its power for achievement without conscience.

The authoritarian conservative character of the Republican party is described in detail by linguist George Lakoff in his book, *Moral Politics: How Liberals and Conservatives Think*. Clues are gained by studying the framework of the teaching of James Dobson of *Focus on the Family* where he presents the family model of the strict, authoritarian father and submissive wife and children. This is presented as God's instruction through the teachings of the Bible. In contrast is the nurturing parent model which is that of the more liberal family approach. One can learn the power of these models by understanding the framework in which they are given and the metaphors used mostly in speeches which are used to express their essences (Lakoff 2002).

Indeed, with the George W. Bush administration, conservatism underwent a significant overhaul. As conservative pundit Fred Barnes described President Bush, as being just what he said he was, "a different kind of Republican," meaning that he presented a different kind of conservatism. The usual conservative view is Jeffersonian in regard to less government, but Bush was more in the school of Madison seeing government as a valuable tool to achieve security, prosperity, and the common good. He used government to achieve conservative ends. Whereas federal governmental power is usually preferred over that of the states, Bush disagreed except in the areas of education and health care. He did not agree with the notion of primarily having the economy finance the government. He saw the first priority of government as national security followed by government's nurturing economic conditions to assure prosperity. In the matter of national security, following the terrorists attack on September 11, 2001, the President declared that America was at war—

an actual war on terrorism. However, he changed the usual approach of responding to the threat as a matter for law enforcement to a new policy of preemptive, or preventive, war by the military, beginning first in Afghanistan in pursuit of the attackers followed by an invasion of Iraq to prevent the possibility of their attacking the United States. This was followed by a series of national interior political reforms both through Congress and by directives of the President as Commander-in-Chief.

George W. Bush was very outspoken about his religious faith, certainly aligning himself mostly with the evangelical wing of Protestantism. One place where his faith played out was in a distinct spiritual tone that could be detected throughout the mission to combat terrorism and the intent to spread democracy wherever God leads. For example, in the themes based on Bush's concept of freedom, liberty, and democracy, terms which often were used interchangeably, he stated that America does not bestow democracy, God does. His faith had an enormous impact on his policies as seen in his support for a faith-based initiative, which he saw as a revolutionary approach to social services by tapping successful private programs and downgrading government efforts. In this he pushed conservatism in a new direction as government could be used as a tool to effect change in partnership with the private sector. Another area in which his faith had a strong influence was in his concern for the sanctity of life. In that vein he opposed the expansion of federal funding for embryonic stem cell research which he thought would inevitably lead to the therapeutic cloning of embryos in exploiting stem cells. He believed that this is creating life only to kill it.

Barnes stated in 2006 toward the end of Bush's second term, "Make no mistake, Bush is redefining conservatism for a new era, consciously moving away from certain precepts that have traditionally characterized the conservative movement" (Barnes 2006, 202). His championing of "results-oriented" government was a significant departure from the view of federal government as having no role. In the meantime Bush brought about a new conservative majority to this country.

From the opposing side of the political spectrum comes a summary statement of current conditions. In a speech cited by Kevin Phillips, Bill Moyers said: "One of the biggest changes in politics in my lifetime is that the delusional is no longer marginal. It has come in from the fringe, to sit in the seat of power in the Oval Office and in Congress. For the first time in our history, ideology and theology hold a monopoly of power in Washington" (Phillips 2004, 218). Phillips, a former Republican strategist, concludes that what we are now experiencing mirrors the excesses of fundamentalism of American, Israeli, and radical Islamic types. It is millennial in character with its emphasis upon the rapture, the end-times and Armageddon. It goes about as far in the direction of theocracy as it can when a political leader claims to speak for God, supported by a ruling party that represents religious true believers, attempting to convince others that government should be guided by religion and implemented into domestic and international political agendas that seem to be driven by religious motivations and scriptural worldviews. Also from within the Republican ranks former senator John Danforth has called for a lessening of the ties to the extreme poles in both parties, but especially for Republicans to move from the religious ideology of the far right that is dominated by radical evangelicalism and fundamentalism (Danforth 2006).

The Call to a Dynamic Middle

What we have described has been largely the push of the extreme poles of the political parties in the sweep of American history. But what about the middle ground? What about the millions of people who do not buy into either liberal or conservative extremes? Robert Wuthnow suggests that there are legitimate roles to be played by the poles in that they define the issues in terms of clear oppositions that set the outer limits of debate and unmask any false consensus that may inhibit an honest rethinking of public values. This leads then to the difficult work at the center which takes place in community where choices are brought together in faithful and creative ways (Wuthnow 1989).

More toward that center, Robert Incahusti (2005) contends that after World War II Americans had come to regard their country more as an economic and military power than as a moral and metaphysical experiment, with this leading to a crisis of personal meaning in the lives of our young. Political collective decision making has become an esoteric art of mass manipulation and control. Yet there is the potential of a spiritual movement within human culture at its best to provide a sustained resistance to the ever-changing face of depersonalization and false authority, a challenge to the complacencies of the middle class, the entitlements of the rich, and the internalized powerlessness of the poor. A giant step in that direction might be to look to the example of the Founding Fathers who simply assumed that democratic institutions existed within a more transcendent cosmic order. How one related to that transcendent order could neither be directed by a nationalist agenda nor curtailed by institutional forces. By refusing to dictate the terms of religious expression the government actually increased the value of a morally engaged, spiritual life, transforming a life lived in accord with conscience into a more liberating and demanding challenge of integrity. With society protected from the tyranny of both majorities and minorities, citizens would be free to direct their religious idealism toward practical human achievements and the quest for a universal ethic.

At the religious and political center there is a stirring among those not content with present directions in the public arena. From within the evangelical sphere there are those who are displeased with the moral state of affairs and are not content to let far right evangelicals and fundamentalists speak for them. Jim Wallis, editor of the magazine *Sojourners*, in his popular book, *God's Politics: Why the Right Gets It Wrong and the Left Doesn't Get It*, has mounted opposition to the positions of both the political left and right by defining a middle position in American politics for dialogue and action. He is an outspoken activist for care of the poor and disenfranchised in our society, a segment that he contends is being largely overlooked and neglected by current social policies. He states that in a world of conflict, war is becoming the primary instrument of foreign policy. Solutions are needed that go beyond the polarized ideological agendas of partisan politics and embrace a political will of bipartisan commitment to a nonpartisan cause both nationally and worldwide.

Wallis states that the spiritual component in all of this is absolutely crucial as it understands how sacred is the blessing of life that undergirds all efforts for justice and peace. His movement is dedicated to offering alternatives toward solutions—ideas and actions, dialogue and collegiality, ecumenical and inter-faith. Applying spiritual values to politics will be the key as he points to the fact that many of the progressive social movements in American history—anti-slavery, women's suffrage, the fight for child labor laws, and the civil rights movement—had overt religious roots and motivations (Wallis 2005). To this Samuel K. Roberts adds that American churches have long taken seriously their role of interpreting and championing spiritual values and moral judgments in accordance with God's intent for human society. This is especially so in the area of civil rights and the quest for social justice, as the issue of race and the reluctance to accord African Americans their full civil rights is an ugly and shameful part this nation's history (Roberts 2000).

Rather than the theological problem of the religious Right, Wallis contends that the big problem is when an administration confuses the identity of the nation with the church and God's purposes with the mission of American empire. What he celebrates is the fact that now moral values are a permanent part of any political agenda. (For good examples of the caliber of the moral values debate see William J. Bennett's *The Book of Virtues: A Treasury of Great Moral Stories* and Jimmy Carter's *Our Endangered Values: America's Moral Crisis*. For more theological renderings of the evangelical shifts see Ronald J. Sider's, *The Scandal of Evangelical Politics: Why Are Christians Missing the Chance to Really Change the World?* and David P. Gushee's, *The Future of Faith in American Politics: The Public Witness of the Evangelical Center*). David G. Myers points to a groundswell of public concern that is visible not just in several public events organized around moral themes, but also broad-based, shared concern for the social ecology that nurtures our children and youth. "The dialogue about American values has shifted from expanding personal rights to enhancing communal civility, from raising self-esteem to rousing social responsibility, from 'whose values?' to 'our values' "(Myers 2000, xi). The more talk about values in political campaigns the better because religion is a primary source of values for many Americans. Wallis calls for "prophetic politics" which finds its center in fundamental moral issues like children, diversity, family, community, citizenship, and ethics. Others could be added, such as nonviolence, tolerance and fairness. What he hopes to accomplish is the construction of national directions that many people across the political spectrum can agree upon. What lies under the strong ethical concern for social issues such as abortion is a desire for "a consistent ethic of life" (Wallis 2005).

This call for dialogue and bipartisan cooperation comes at a time when Robert D. Putnam says there is a growing distrust of public institutions and government. This may be due in part to the broad and continuing erosion of civic engagement that began a quarter-century ago. Add to that the current trends toward polarization and skepticism, toward government and institutional leadership, and one becomes increasingly aware that democracy depends to a large degree upon the existence of a vibrant "civil society" (Putnam 2000). Now, as our government desires to plant democracy in the Middle East and in other places in the world at large, it is important to understand that many people question that wisdom as they raise the question: how can you plant and grow vibrant civic life in soils traditionally inhospitable to self-government?

A move toward a dynamic middle is called for as we see emerging resistance to the extremes of the far right of political and religious conservatives from the political parties and general society at large. Leadership toward that dynamism may come from those voices of reason within the realm of sociological and theological fields of study. Certainly political debate and activism will provide dynamism from the likes of Jim Wallis and Sojourners who act from that evangelical base to address moral issues of our society. To the essence of that movement let us add the suggestion of communitarian Professor Amitai Etzioni that we embrace a new Golden Rule—not that we want to replace the old one. While blending virtues of tradition with the liberations of modernity, he suggests that we recapture the spirit of civil community and regenerate the social order by following the new golden rule: "Respect and uphold society's moral order as you would have society respect and uphold your autonomy" (Etzioni 1996, xviii).

Future Trends—2020

What can we envision for the political arena, socially and spiritually, as we look toward the year 2020? As a person of faith with a keen interest in community and the politics that go along with it, I would venture a few projections—not predictions.

The first is the belief that spirituality will increase as a reality in intensity and breadth. The world is growing smaller and flatter (Friedman 2005) through technological advances in communications, increased ease of travel, and economic growth and shifts. In such a world the forces of the Third World will continue to reach for the next rung on the ladder that lifts them toward the level on which the Western World has enjoyed prominence and affluence. The clash of ideologies, especially with religious extremes of fundamentalism, will increase until the majority of persons of moderate persuasion assert their leadership to bring about a more temperate spiritual and religious climate and a more moderate political one as well. Wars as symbol and tactic may continue, but perhaps they will be more territorial and economically based than religious. The United States will be less eager to spread democracy by force and more through persuasion and diplomacy, especially as other democracies move to the forefront in a leadership role to supplement and present an alternative model of democracy along with the U.S. model.

In the United States, spirituality will continue to include expressions of a variety of religions. As we continue to become more pluralistic we will move more toward an interfaith complexity of mutual respect. Catholicism and Protestantism will continue to dominate for several more generations but diminish in time mainly due to United States population's growth through immigration. All religions will be challenged by technological developments that will intensify debate over moral values and issues with extreme voices continuing in their role of moral critic for the greater society. Hopefully strident attempts to dominate society through political imposition of morality will be tempered with reason and respect and by the practice of more civil discourse among parties. Furthermore, a curb on lobbyist and campaign spending will be needed. Also, somehow, a major attitude change is needed in elected officials—a change of their first priority as being reelected, to an attitude of service with integrity.

More pointedly, in politics we will witness the swing of the pendulum back toward the middle away from the far right of conservatism that has been led by neoconservatives in league with religious evangelical conservatives and fundamentalists. Both elements will lose some power due to a decline in numbers, a measure of success in addressing certain issues, and the resistance from the rejuvenated new center and left of the political sphere. However, the center will be more to the right than the center position of the twentieth century. As Robert Wuthnow optimistically projects, hopefully the religious Right will be less concerned with achieving its ends through political means alone, and will be more devoted to the ideals of service, caring for the poor and disadvantaged of society, promoting community, reconciliation, and the transmission of values through teaching and training the young (Wuthrow 1989).

In the short term, as David M. Abshire declares, we will continue to be challenged by a list of issues that include at least the following: global war on terrorism, conflicts in Iraq, Afghanistan, and the Middle East, the solvency of Social Security and Medicare, the national deficit, the K-12 education crisis, widespread anti-Americanism, and the erosion of character-based leadership among the clergy, schools, heads of business, and political leaders, including the presidency. To Abshire's list one may add the growing concern about global warming or climate control and our nation's dependency upon fossil fuels for energy. These call for approaches marked by civility which means respect, listening, and dialogue as we attempt to move to a higher common ground (Abshire 2005). In the longer term, today's moral issues will linger for some time on our national agenda, and every one of them will be debated on moral and ethical grounds. As the heretofore quiet middle is awakened, compromises will be reached on major life issues such as abortion, capital punishment, euthanasia, and stem cell research. At the same time a warning is sounded by David G. Myers regarding our socio-economic divide. There is the grave possibility that more future conflicts may erupt between rich and poor than between ethnic groups and nationalities. "In a society where the top one percent hold assets equal to those of the bottom forty percent, talk of revolt may flourish when the next recession hits" (Myers 2000, 143ff). However, the poverty that plagues us, he says, is not only an absolute material poverty, but also a relative poverty that breeds spiritual starvation. If we believe that all is hopeless, then that is what will be. Yet ". . . [I]f we can agree that prosperity must be seasoned with purpose, capital with compassion, and enterprise with equity, then maybe the best is yet to come" (2000, 143ff).

The attack of September 11, 2001, and the subsequent wars in Afghanistan and Iraq have convinced us that there are no longer chasms deep enough or walls high enough to give us protection from others or to protect them from us. Consequently we might, as Canon Charles Gibbs suggests, begin to see ourselves as citizens of the earth and children of the abiding Mystery at the heart of all that is. "Then, with open hearts and appreciative, inquiring minds, set out on a journey to encounter the other and find ourselves" (Gibbs 2005, 229).

References

Abshire, David M. 2005. "The Grace and Power of Civility: Commitment and Tolerance," *Deepening the American Dream: Reflections on the Inner Life and Spirit in Democracy*, Mark Nepo, ed. San Francisco: Jossey-Bass.

Aburdene, Patricia. 2005. *Megatrends 2010: The Rise of Conscious Capitalism*. Charlottesville: Hampton Roads Publishing Company.

Barker, Joel A. and Scott W. Erickson. 2005. *Five Regions of the Future: Preparing Your Business For Tomorrow's Technological Revolution*. New York: The Penguin Group.

Barnes, Fred. 2006. *Rebel-in-Chief: Inside the Bold and Controversial Presidency of George W. Bush*. New York: Crown Forum.

Bella, Robert, Richard Madsen, William M. Sullivan, Ann Swidler, and Steven M. Tipton. 1996. *Habits of the Heart: Individualism and Commitment in American Life*. Berkeley: University of California Press.

Bennett, William J., ed.1993. *The Book of Virtues: A Treasury of Great Moral Stories*. New York: Simon & Schuster.

Carter, Jimmy. 2005. *Our Endangered Values: America's Moral Crisis*. New York: Simon & Schuster.

Danforth, John. 2006. *Faith and Politics: How the "Moral Values" Debate Divides America and How to Move Forward Together*. New York: Penguin Group.

Dean, John W. 2006. *Conservatives Without Conscience*. New York: Penguin Group.

Etzioni, Amitai. 1996. *The New Golden Rule: Community and Morality in Democratic Society*. New York: HarperCollins.

Friedman, Thomas L. 2005. *The World Is Flat: A Brief History of the Twenty-first Century*. New York: Farrar, Straus and Giroux.

Gibbs, Charles. 2005. "Opening the Dream: Beyond the Limits of Otherness," *Deepening the American Dream: Reflections on the Inner Life and Spirit in Democracy*. Mark Nepo, ed. San Francisco: Jossey-Bass.

Gushee, David P. 2008. *The Future of Faith in American Politics: The Public Witness of the Evangelical Center*. Waco: Baylor University Press.

Hastey, Stan. 2000. "Reconstructing the Revolution: Religion's Rightful Role." *Review and Expositor*, Winter 2000. Louisville: Review and Expositor.

Inchausti, Robert. 2005. "Breaking the Cultural Trance: Insight and Vision in America," in *Deepening the American Dream: Reflections on the Inner Life and Spirit in Democracy*, Mark Nepo, ed. San Francisco: Jossey-Bass.

Johnston, Whittle. 1981. "Ethics, Power, and U.S. Foreign Policy," in *Christianity and Politics: Catholic and Protestant Perspectives*, Carol Friedley Griffith, ed. Washington, DC: Ethics and Public Policy Center.

Lakoff, George. 2002. *Moral Politics: How Liberals and Conservatives Think*. Chicago: University of Chicago Press.

Lindorff, Dave and Barbara Olshansky. 2006. *The Case for Impeachment: The Legal Argument for Removing President George W. Bush from Office*. New York: St. Martin's Press.

Myers, David G. 2000. *The American Paradox: Spiritual Hunger in An Age of Plenty*. New Haven: Yale University Press.

Napolitano, Andrew P. 2006. *The Constitution in Exile: How the Federal Government has Seized Power by Rewriting the Supreme Law of the Land*. Nashville, TN: Nelson Current.

Newberg, Andres, Eugene d'Aquili, and Vince Rause. 2001. *Why God Won't Go Away: Brain Science and the Biology of Belief*. New York: Ballantine Books.

Neuhaus, Richard John. 1981. "The Post-Secular Task of the Churches," *Christianity and Politics: Catholic and Protestant Perspectives*. Washington, DC: Ethics and Public Policy Center.

Pagels, Elaine H. 2005. "Created Equal: Exclusion and Inclusion in the American Dream," in *Deepening the American Dream: Reflections on the Inner Life and Spirit of Democracy*, Mark Nepo, ed. San Francisco, CA: Jossey-Bass.

Phillips, Kevin. 2006. *American Theocracy: The Peril and Politics of Radical Religion, Oil, and Borrowed Money in the 21st Century*. New York: Penguin Group.

Putnam, Robert D. 2000. *Bowling Alone: The Collapse and Revival of American Community*. New York: Simon & Schuster.

Ray, Paul H. and Sherry Ruth Anderson. 2000. *The Cultural Creatives: How 50 Million People Are Changing the World*. New York: Three Rivers Press.

Roberts, Samuel K. 2000. "Marching to Zion: American Churches and the Struggle for Civil Rights and Social Justice." *Review and Expositor*. Winter 2000. Louisville: Review and Expositor.

Roof, Wade Clark. 1999. *Spiritual Marketplace: Baby Boomers and the Remaking of American Religion*. Princeton: Princeton University Press.

Sider, Ronald J. 2008. *The Scandal of Evangelical Politics: Why Are Christians Missing The Chance to Really Change the World?* Grand Rapids: Baker Books.

Tillich, Paul. 1963. *The Eternal Now*. New York: Charles Scribner's Sons.

Wilber, Ken. 2005. *A Sociable God: Toward a New Understanding of Religion*. Boston, MA: Shambhala Publications.

Wuthnow, Robert. 1989. *The Struggle for America's Soul: Evangelicals, Liberals, and Secularism*. Grand Rapids: Eerdmans.
Wuthnow, Robert. 1993. *Christianity in the 21st Century: Reflections on the Challenges Ahead*. New York: Oxford University Press.
Zohar, Danah and Ian Marshall. 2000. *S Q: Connecting With Our Spiritual Intelligence*. New York: Bloomsbury Publishing.
Zohar, Danah and Ian Marshall. 2004. *Spiritual Capital: Wealth We Can Live By*. San Francisco: Berrett-Koehler Publishers.

Chapter 5

Economics

ROBERT STRAYER AND JANICE MICHAEL

> It goes without saying that we will not be able to buy world peace in the supermarkets. We will not be able to eliminate economic discrimination just by pouring more money into the economy. A world at peace and free of discrimination begins in our inner soul.
>
> (Sfeir-Younis 2004, 3)

Historical Perspective

One of the unchanging realities of human existence is that while people and their societies have unlimited needs and wants, their resources to satisfy those needs and wants are limited. Based on the Greek words *oikos* (house) and *nemein* (to manage), economics is the study of the systems people develop to deal with the problem of scarcity in order to determine what to produce, how to produce, and who is to receive the output that is produced in the society. The fact that wants/needs are unlimited and resources are finite forces the society and its people to make decisions between competing choices.

The model that economists use to illustrate this universal problem is a production possibilities curve. Many times this model is used in explaining public policy as a "guns and butter" model. In this simplistic version, a society must choose between using its resources for production of defense (public goods) and devoting resources to consumer private goods. If all resources are used for production of goods for defense without any production of consumer goods, the society will lack food, shelter, and other consumer goods. If all resources are devoted to food, shelter, and consumer goods production and none to defense, then the society leaves itself open to invasion. Either extreme is unacceptable, so the society tries to produce some of each possibility to balance meeting the needs of its citizens.

The one thing that makes the decision process difficult is that choosing one outcome over the other involves increased sacrifice of production of the alternative choice because resources are not interchangeable in use. For instance, to convert a factory from the production of military tanks to the production of public transit busses or private automobiles involves time and money to retool and reorganize production. The time lost cannot be recovered and the increased cost of conversion at some point may become prohibitive. This opportunity cost, the explicit and implicit cost of changing resource usage, becomes an "opportunity lost" because once the decision has been made, the competing choice is no longer an option. It really is true that one "cannot have one's cake and eat it, too." Without any increase in resources or a change in the technology for using the resources, a society experiences no economic growth.

Economic activity requires organization and interaction. One way to illustrate this is the concept of an "economic wheel" that suggests the relationships among domestic businesses, consumers, government, and foreigners. Without businesses, we would be nomads trying to supply our basic physical needs. The earth could not support us. Businesses supply goods and services to consumers, other businesses, governments, and foreigners. In return for the goods and services, businesses receive money which functions as a medium of exchange. Businesses use money to secure the resources (also called "the factors of production" needed to continue providing goods and services. The money paid to those holding the resources becomes income for them.

Resources fall into four categories: land, labor, capital, and entrepreneurship. Those who supply land or natural resources receive an income called *rent*. The suppliers of labor receive an income called *wages*. The supplier of capital, stored land, and labor receives an income called *interest*. Entrepreneurs, the risk takers, receive an income called *profit*. These suppliers of resources compete with each other for a piece of the economic pie. If one group gets a bigger piece, the other groups gets smaller pieces. For example: if the price of oil increases, those who use oil or products derived from oil will be paying more and so will have less ability to purchase other goods and services; this can result in a lower standard of living. If, in an attempt to deal with the higher prices, wage earners ask for and receive a higher wage, other suppliers will withhold resources from the market until a new balance is achieved.

While all organized economies face this same problem, what differentiates economies is how they organize to make the decision of what to produce, how to produce, and how to distribute the production. Economists have generally grouped economic organization into three general categories: traditional, market, and command. In a traditional economy, the decisions are made in the manner that they have always been made. One of the best known examples within American society of a traditional economy is the Amish community. The religious beliefs of the community dictate that its members farm or produce goods for sale to the outside world using the same technologies their grandparents and great-grandparents used. If and when a new technology is introduced, it must pass the test of complying with the religious beliefs before being accepted.

The other two organizational forms, command economy (communism and socialism) and market economy (capitalism), are often thought of as opposite ends of the organization spectrum. While neither has ever existed in its purest form, they have existed in forms that are close enough to be able to generalize characteristics of each. In the command economy, the decisions about production and distribution are made by a central decision making body and a system of committees. In both communism and socialism the capital goods (those goods used to produce other goods and services), are owned by the central government rather than private individuals or companies. The final goods and services are also publicly held in the communist economy, while they are privately held in a socialist economy. While distribution processes may seem well-organized when reviewed on paper, in practice they are often inefficient with insufficient production and poor quality goods and services as a result. One of the few remaining examples of a command economy is North Korea. While independent, reliable data is hard to find, according to *The 2007 World Factbook*, published by the CIA, the North Korean economy is one of the poorest in the world because of the emphasis on military production over meeting the basic needs of citizens. By the end of 2006, the country was facing the twelfth straight year of food shortages due to the combined forces of misallocation of resources and weather extremes of flooding and drought in consecutive years. While the central government did allow humanitarian aid to come into the country to avoid mass starvation, those efforts were curtailed in 2005 and North Korean citizens remain among the poorest in the world with an estimated yearly income of $1,800 per capita (CIA 2007).

The polar opposite of the command economy is the market or capitalist economy where both capital and consumer goods (those goods used to provide personal satisfaction) are privately held. The fact that they are privately held means that there is also the right to exclude others from their use or enjoyment. Adam Smith, the Scottish economist and "father of modern economics," was one of the first to formalize the concept of a market economy in his work, *An Inquiry into the Nature and Causes of the Wealth of Nations* (Smith 1776). In a market economy, the three economic questions of what to produce, how to produce, and for whom to produce, are answered by a system of prices determined by the interaction of the forces of supply and demand. Private ownership of resources and self-interest to gain the most from use of those resources motivate production decisions. He argued that a nation's wealth is created by the ability of its people to produce goods and services and trade in free markets. He envisioned that the role of government should be a very conservative one of providing national defense, education, infrastructure, and a system of courts to enforce contracts. He hypothesized that a system of markets acts as an "invisible hand" to assure that the best interests of society are served when producers compete with one another to win consumers. Consumers "vote with their dollars," rewarding producers for making quality products, spurring innovation and creativity, and providing a wide range of goods and services (Smith 1776).

Smith based his work on observation of the changes taking place as the Industrial Revolution was unfolding; the system of markets as he envisioned them, while not working flawlessly, did work and the Western economies based on his theory prospered. Specialization of production led to mass production of goods which allowed entrepreneurs to build individual wealth, thus providing capital for expansion of economic activity. Throughout the eighteenth and nineteenth centuries, it was accepted that "the business of business is business," and most western economies functioned with a "laissez faire" or "hands off" role for government. While there were social critics (Charles Dickens and his venerable work, *A Christmas Carol*, comes to mind), the prevalent and accepted theory of how a modern industrial economy functions was the capitalist system described in Smith's theory.

It was assumed that while the economy may experience periods of recession, the flexibility of wages and prices determined by supply and demand forces would allow the economy to be self-correcting and prosperity would return. The role of government was limited; government involvement would upset the balance of interaction between business and consumers, so it was important for government to maintain a balanced budget so that taxes collected were returned to the economy in the form of government spending. Consumption by individuals and investment by businesses were the source of economic activity and growth. Income inequality was a given; motivated and hard-working individuals would reap the rewards of their efforts. Those with less ambition or skills would enjoy less wealth. The least fortunate would be cared for by their families or charitable institutions such as churches. During the 1880s, a number of Protestant denominations embraced a religious philosophy known as the Social Gospel. Church members were exhorted to improve living conditions as well as evangelizing into the faith. One of the most prominent Social Gospel leaders was a Congregational minister and writer, Washington Gladden. He linked Christianity with active social service to provide hospitals, schools, orphanages, and homes for the aged who had no families to care for them. In urban centers there was an organized movement to call attention to the plight of the poor and to eradicate the conditions contributing to poverty, most notably Jane Addams and her Chicago institution, Hull House (Divine, et al. 2005).

While their efforts did raise social consciousness, the issues of poverty, unemployment, and income inequality were still generally viewed as being outside the realm of government responsibility. Government was involved in the economy, but it was still in the traditional role of providing a climate conducive to the conduct of commerce such as a standard currency and tariffs to regulate trade. The last half of the nineteenth century was the era of corporate and personal fortunes made in oil, J. P. Getty, J. D. Rockefeller; in railroads, E. H. Harriman, George Pullman; in banking, Andrew Mellon, J. P. Morgan; and in the steel industry, Andrew Carnegie. Political pressure to curb the abuse of economic power led to passage of such landmark legislation as the Sherman Anti-Trust Act of 1890 which declared monopolistic control to be an obstruction to free market activity.

As the American economy became more industrialized, there was an increase in the effort to organize workers into unions. Labor unions had existed even in colonial times, but they were organized according to trades and craftsmanship. The industrial union which sought to organize all workers in an industry rather than by skills was met with legislative and judicial resistance. The Supreme Court upheld the use of the court injunction to break the Pullman Railroad Strike of 1894 (Devine, et al. 2005). One of the first applications of the Sherman Anti-Trust Act was its use against organized labor, declaring unionization an illegal restraint of trade. It wasn't until the passage of the Wagner Act in 1935 that workers were guaranteed the right to form unions and to engage in collective bargaining.

In the half century after the Civil War, the American economy was transformed from a rural, agrarian-based economy to an industrialized, urban one. In 1860, the national output was estimated to be worth $16 billion; by 1900 it had risen to $88 billion. In the first two decades of the twentieth century, innovations such as the mass-produced automobile, advances in aviation, motion pictures, radio, and an increasing array of affordable consumer goods, all changed the structure of the American economy (Devine et al. 2005).

Throughout the eighteenth and nineteenth centuries, the economy had experienced periods of alternating levels of economic activity with periods of higher levels of activity (peak) followed by falling activity (recession), a lowest point of activity (trough) and a resurgence of activity(recovery). While the timing and duration of each phase of the business cycle was not predictable, market activity always seemed to move the economy on through the cycle to the next phase. With the stock market crash in October 1929 and the onset of the Great Depression, economic activity plunged. The Gross National Product (GNP), the measure of the economic growth, went from growing at roughly three percent per year to shrinking by nearly fifteen percent from 1929 to 1930 (Bureau of Economic Analysis 2004).

As the Depression transformed the demographics of the country, economic theory also underwent a transformation. Classical market economy forces described by Adam Smith seemed to be ineffective in dealing with the downward-spiraling economic conditions. The Hoover administration practiced the traditional role defined for government (maintaining a balanced budget while providing a climate conducive to economic activity) by passing legislation to increase tariffs to protect domestic jobs, as well as raising taxes and cutting spending to bring the budget into balance. The Depression deepened; parents who could no longer provide for their children placed them in orphanages or divided them among extended family members in other areas where work was still available; the desperation described by John Steinbeck in *The Grapes of Wrath* afflicted thousands of Americans. Increasing numbers of unemployed workers who had lost everything they owned congregated in makeshift shack towns dubbed "Hoovervilles" (Garraty 1966, 728). In the 1932 presidential race, challenger Franklin Roosevelt promised a "New Deal" to put the economy on the road to recovery and alleviate the twenty-five percent unemployment rate.

Part of the New Deal was experimenting with an economic theory proposed by John Maynard Keynes, a professor at Cambridge University in England. In his work *The General Theory of Employment, Interest, and Money,* Keynes (1936) contended that wages and prices were not as flexible as had been believed and there was no assurance that the economy would be self-correcting, even over the long run. He contended that the situation was so dire that governments could not wait for the long run, and, rather than put emphasis on a balanced budget, governments needed to take an active role in the economy to stimulate recovery. He proposed the concept of aggregate (total) demand in the economy consisting of four components: consumer spending, business investment, government spending, and net exports (exports minus imports). Classical thinking also identified these sectors of economic activity, but what was different between Keynes' approach and the classical approach was the relationship among the elements.

In classical theory, wage and price flexibility would allow businesses and consumers (providers) of the factors of production to interact so that temporary imbalances would work out according to the principle of supply and demand. A favorable balance of trade, with exports exceeding imports, would increase employment while the goods and services being produced domestically and sold abroad would further enhance economic conditions. The two places where money could leave the system and create an imbalance were in taxation and savings. Classical theory assumed that banking would operate on the principles of supply and demand: private individuals would save by putting money on deposit in the banking system, and then businesses would borrow through the banking system and thus return the funds to the economy. If private savings increased, the interest rate would fall and businesses would increase borrowing; if savings decreased and there was less money in the banking system, interest rates would increase and fewer businesses would borrow existing funds. With this sector also being self-correcting, it was important for government to keep a balanced budget so that any money levied in taxes was returned to the economy in the form of government spending.

Another critical underpinning was the theoretical work of Jean Baptiste Say, which to the general public is commonly referred to as "Say's Law." Say's Law can be stated briefly: "Supply creates its own demand." In other words, if goods and services are being produced, those workers producing them will be earning an income that will, in turn, allow them to purchase the goods and services being created.

Keynes argued that Say's Law was not necessarily true; that instead of the Depression being due to an economy that was out of balance, the economy was, in fact, stable and in balance, but at a point well below full employment. Workers who were employed were earning a wage, but it was not sufficient to purchase enough goods and services to result in economic growth. If that were true and the economy stable, then the banking system could not restore a higher level of economic activity. Keynes theorized that this was especially true because consumers have one set of motives for saving that would make money available for investment loans, while businesses have a different set of motives for borrowing their saved funds. Individuals need to have a sufficient level of income in order to be able to save. Keynes proposed that individuals have three reasons for saving money. The first is to have funds for performing transactions. They save in order to purchase something of higher value than what could be purchased out of regular income. The second reason was to prepare for unexpected major expenses—the proverbial "rainy day". The last reason to save was to have funds for speculative purposes, something not possible—or desirable—with lower levels of reserve.

Keynes also argued that businesses were not so concerned about an interest rate set by the forces of supply and demand in the money market as they were about the rate of the return earned by capital they borrowed. Businesses would still borrow at a seemingly high interest rate if the rate of return on their investment were still higher; for a similar reason, they might not borrow at a very low interest rate if their anticipated rate of return would be even lower. Once this disconnect occurred between savers and borrowers, the ability of the system to self-correct was no longer assured, particularly when large numbers of banks failed and depositors lost their entire life savings. Keynes argued that when aggregate demand and aggregate supply (the total output of the economy) was stabilized at such a low level of activity, government should take a more active role in financial activities to stimulate growth in the economy. On the other hand, if the economy had stabilized at a level of economic activity that was too high to sustain without inflationary pressures, then government should have actively pursued measures to slow the economic activity, thus lowering the cost of goods and services. Changes in aggregate demand were the driving force behind the business cycle, and rather than government's maintaining a balanced budget, it was more important to adjust fiscal policy, i.e., the way it taxes and then spends the tax revenues to achieve such economic goals as stable prices, economic growth, and full employment. Government budgeting should be planned to counteract the business cycle; while not eliminating the cycle, government could moderate it through its taxing and spending policies. He advocated that during the peaks of economic activity, governments should tax more and spend less, creating a budget surplus to help control inflation. During periods of recession and depression, government should stimulate the economy by taxing less and spending more, even though it produces a budget deficit.

The Roosevelt administration produced a number of programs that were Keynesian-based, most notably the Works Progress Administration (WPA), the Civilian Conservation Corps (CCC), and the Tennessee Valley Authority (TVA). Under the WPA, workers were "given a hand up instead of a handout." Public works projects such as railway terminals, school buildings, post offices, and bridges were built. Artisans were hired to paint murals, create mosaics, and sculpt artwork for the public spaces. The Union Terminals in Cincinnati and Los Angeles are outstanding examples of Art Deco design that were commissioned as WPA projects. The WPA developed the Federal Theater Project which put thousands of actors, directors, and stage crews to work on theater productions which entertained nearly 60 million people. The CCC planted thousands of trees, constructed fire watchtowers in national forests, and built many of the facilities still in use in the national park system. The TVA constructed a series of dams across the Tennessee River Valley that not only helped control flooding, but provided electricity to the region (Garraty 1966).

Funding such major public works projects gave those workers hired an income that could be used to pay taxes, to purchase consumer goods and services, and to save. What was spending for them now became an income for someone else. Keynes labeled this "the multiplier effect" and theorized that any change in spending—whether by individuals as consumers, businesses for investment, or by government in spending on programs, goods, or services—would create a larger "ripple effect" of spending only limited by how much is saved out, held back, and no longer available for spending to create a higher level of activity. He also theorized that tax cuts could stimulate the economy, even without an initial round of government spending.

In addition to the Wagner Act of 1935 which gave labor the right to organize, the other major landmark legislation to come out of the Great Depression was the passage of the Old Age, Survivors and Disability Insurance Act (OASDI): The Social Security Act of 1935. It set up a national system of old-age insurance, a state-federal system of unemployment insurance, and benefits for children under the age of eighteen. The system was designed to be a "pay-as-you-go" system financed by payroll taxes with both workers and employers contributing. Critics of the act argued that it was the first step towards socialism and went against "traditional American characteristic of independent can-do" (Garraty 1966).

Whether or not the application of Keynesian theory was responsible for the end of the Great Depression is debatable; the outbreak of World War II and the increased production needed to meet the war effort pushed the economy to its limits. The American government raised millions of dollars to fund the war effort through the sale of bonds; the purchase of bonds not only provided needed funding beyond what was available through taxation, but also lessened the inflationary pressures on the economy as the incomes being earned were diverted from driving up the prices of scarce rationed commodities such as gas, sugar, and tires.

The real fear at the end of the war as veterans returned home was how to re-absorb them into the civilian economy without throwing that economy back into a depression. Congress passed The Employment Act of 1946 which created a Council of Economic Advisors in the executive branch and codified that the federal government has the responsibility to "use all practical means consistent with free competitive enterprise to create conditions under which all able individuals who are willing to work and seeking work will be afforded useful employment opportunities" (Tucker 2006, 230). However, the transition from wartime production to consumer goods to meet the pent-up demand from the war years and the outbreak of hostilities in the Korean peninsula relieved increasing unemployment pressures.

By 1958, the economy was experiencing recession and unemployment rates had reached six percent. The Eisenhower administration which had been elected by a landslide in 1952 and re-elected in 1956, was coming to a close and had lost congressional seats to the Democrats during the midterm elections. Even though the federal budget had been balanced for three of the most recent six fiscal years, many U.S. citizens were concerned about the nearly four billion dollars going overseas as military and economic aid, primarily to check the spread of communism. With the election of John Kennedy in 1960, there were new ideas and an increasing reliance on a Keynesian approach to the economy, due in large part to his appointments of such academics as John Kenneth Galbraith to the Council of Economic Advisors. In spite of some initial successes in Congress, other proposals such as hospital care for the aged under the aegis of Social Security were defeated (Bailey 1971). With the escalating costs of involvement in Southeast Asia, the national debt reached $296,170,000, nearly $40,000,000 more than at the close of World War II. The economy was still suffering unemployment and at the time of his assassination, Kennedy was working to secure votes for a Keynesian fiscal policy of cutting taxes and increasing federal spending to stimulate the economy. Within only a few months, Congress enacted the legislation Kennedy had been seeking (1971) .

Following a landslide re-election victory over Senator Barry Goldwater, President Lyndon Johnson launched his "Great Society" program in January 1965. One key piece of legislation was the Social Security-provided healthcare program for senior citizens that Kennedy had previously sought—Medicare. While decried as a move toward socialism by both conservative Republicans and Democrats, the program won acceptance from nearly 17 million senior citizens (Bailey 1971). Throughout the administrations of Presidents Nixon, Ford, and Carter, social and income support programs expanded and government involvement burgeoned in economic affairs previously the domain of the private sector. In 1929, federal government expenditures as a percentage of the GDP was four percent; by 1970 it was twenty percent (Economic Report of the President 1975).

At the beginning of the 1970s persistent inflation plagued the American economy. While the end result, higher prices, was the same, two different causes contributed to the problem. Increased government expenditures increased aggregate demand, and sudden oil shortages increased resource prices. Consumer buying power shrank in spite of larger dollar incomes. When workers negotiated contracts with built in cost-of-living allowances, COLAs, the increases were often passed on to consumers in the form of higher prices, setting off another round of demands for higher wages fostering a wage-price spiral. In an effort to combat the wage-price spiral, the Ford administration imposed price controls which did help bring the inflation rate down to one percent in 1973. But once controls were removed, inflation returned to double-digit levels until 1980 (Economic Report of the President 2005).

The election of Ronald Reagan in 1980 marked another turning point in the political application of economic theory in the American economy. Reagan campaigned on a platform to reduce the size of government and its involvement in economic affairs, specifically to reduce the amount of burdensome regulations on business and to cut taxes on both individuals and businesses to stimulate productivity. As a proponent of classical supply-side economics, he argued that incentives such as accelerated depreciation on investment and reduced marginal tax rates would increase aggregate supply. This in turn would increase employment with the output of goods and services lowering price levels. Unlike the Keynesian fiscal policies which focus on too little aggregate demand resulting in unemployment and too much aggregate demand resulting in inflation, supply side policies which increase aggregate supply deliver both higher employment and lower price levels at the same time.

While the economy did experience lower inflation and lower unemployment throughout the Reagan administration, there were mounting concerns over the increasing size of the federal debt. Between 1980 and 1990, persistent deficits pushed the national debt from one trillion dollars to a little over three trillion dollars; as a percentage of the GDP, it went from twenty-five to almost sixty percent of GDP (Economic Report of the President 2005). One of the major concerns over the size of the debt is that in order to continue operating, government must continue to borrow, placing it in direct competition for borrowed funds with businesses and consumers. This is primarily due to the role of the central banking system in the economy.

Since the passage of the Federal Reserve Act of 1913, the central banking system has played the major role in stabilizing the economy by either increasing or limiting the amount of credit available to consumers and businesses. Normally, during an inflationary period government will offer bonds and securities with a higher interest rate to encourage individuals to purchase those rather than contributing to inflation by spending more. During a recessionary period, government will buy back bonds and securities to put more money back into the economy. Banks may also purchase or sell off bonds; if a bank can get a better rate of return by purchasing government securities instead of extending a loan, it will use excess reserves to purchase securities. If government continues to borrow to finance increasing debt, consumers and businesses are "crowded out" of the market for loanable funds. Critics of this position would argue that unless the economy is at full employment, crowding out is not as likely to happen as when the economy is at less-than-full employment.

The combination of federal spending, federal deficits, and the national debt are likely to continue to be a source of considerable debate. In 1970, income security programs were twenty-two percent of the budget while defense spending was forty percent. (Economic Report of the President 2005). By 2006, income security programs were forty percent of the budget while defense expenditures had fallen to nineteen percent, in spite of increased expenditures for the war on terrorism (Economic Report of the President 2007). With the exception of 1998-2002, when there were rising budget surpluses, the federal budget has generated deficits ranging from a $50 billion deficit in 1970 to a $400 billion deficit in 2004 (Economic Report of the President 2005). By February 2007, the national debt was $8.8 trillion; with deficits projected to continue well past 2010, the debt is also projected to increase (Congressional Budget Office, Budget Outlook 2007).

Future Concerns

The Increasing Share of Federal Budget on Social Spending and Growth of Government

The goals of high levels of employment, stable prices, and economic growth have been the focus of national economic policy since the Great Depression of 1930. Prior to the Great Depression, federal involvement in the economy was limited primarily to defense spending and very little spending that would be viewed as "social spending" on programs or direct payments to individuals. With the Great Depression and the passage of The Social Security Act, the process of institutionalizing social spending had begun. In subsequent years, federal entitlement programs have become an increasingly larger part of the federal budget.

The Congressional Budget Office is projecting that in 2010, the increase in the number of baby-boomers retiring will increase the annual growth rate of Social Security spending. While the projected growth rate for Social Security spending is near four and one-half percent in 2008, the rate of increase will be six and one-half by 2017 (Congressional Budget Office, Budget Outlook 2007).

Two health care components of the social spending portion of the federal budget, Medicare and Medicaid, are nearly as large as the spending on Social Security itself. In 2006, the federal government spent $549 billion on Social Security; Medicare spending was $330 billion and Medicaid spending was $191 billion (Economic Report of the President 2007). The Congressional Budget Office projects that each of these programs will grow in the range of 7-8% annually with total outlays for the programs more than doubling by 2017 while the GDP is projected to only grow by three-to-four percent annually. Assuming no change in current taxation structures or reduction of benefits, by 2017 spending on Social Security, Medicare, and Medicaid will equal nearly eleven percent of the GDP, up almost two percent from 2007. If current law remains in place after 2017, spending for health care is likely to rise to more than twenty percent by 2050 unless projections of economic growth are too low (Congressional Budget Office 2007).

There is considerable debate over the increasing use of federal dollars for providing health care and how much that use contributes to increasing health care costs. According to the Bureau of the Census, in the United States, one out of every seven dollars spent is spent on health care, a greater percentage than any other industrialized country (Bureau of the Census 2004). In the absence of insurance, private or public, the price of health care is determined by the interaction of supply and demand. With the introduction of public and private insurance, consumers no longer bear the real cost of the service. Because they perceive the price to be lower than it actually is, they increase the quantity they demand. In order to meet the quantity demanded, health care providers increase the quantity of health care provided, but at a cost that is higher than what would be determined by market forces. Rather than resulting in a shortage of health care, which would happen in the absence of insurance, the fact that insurance pays the difference keeps the care available, but at a higher cost.

The factors, other than price, that can change the demand for normal goods also affect the market for health care. Heavy advertising for new drug therapies, an aging population, public subsidies for health care for low income persons, and increases in tobacco, alcohol, and drug abuse all can increase the demand for health care. Changes in consumer attitudes toward healthcare, particularly for elective procedures such as cosmetic surgery, also increase demand. As demand increases, price will increase. Health care providers will then increase the quantity of health care available in response to the higher price.

In a normal market situation, there would be an increase in supply which would help lower the price of the good or service. However, in the health care system the complex regulations that exist to ensure the safety and quality of the health care actually prevent the increase of supply, so prices remain higher than they would in the absence of insurance (Tucker 2006).

Critics of federal social spending would argue that the growth of social spending to forty percent of the budget has diminished personal responsibility and pushed the American economy closer to socialism. They would point to the increase in social spending and the increase in out-of-wedlock births as one trend to support this claim. In 1940, 38 of every thousand births were to single mothers; in 1970 that number had increased to 107; in 2000, to 332; by 2004, to 358. Preliminary data suggest this trend will continue (National Centers for Disease Control 2004). In many cases, it is the grandparents who end up raising the grandchildren. According to the U.S. Census Bureau, in 1970, 2.2 million children or a little over three percent of children under the age of eighteen were living in homes maintained by their grandparents. In 1997, the number had risen to five and one-half percent or 3.9 million children (Bryson and Casper 1999). By 2002, the number of children living with grandparents was 5.6 million, eight percent of all children. Of those living in a home where the grandparent was the homeowner and no parents were present, seventeen percent were receiving public assistance (Fields 2003).

Critics would also contend that personal property rights have been increasingly eroded due to increased government regulation and unfunded mandates. For example: with federal agencies such as the Food and Drug Administration (FDA) controlling the testing process that delays introduction of new medications into the market place, critics argue that the regulatory climate infringes on the private property rights of the company owners; rather than provide a climate conducive for commerce, government hinders commerce. Private businesses must meet numerous regulations to meet standards for disability access, fair labor standards, and environmental impact, but there are no funds to compensate them for doing so.

The general increase in social spending programs of all types is a source of controversy. The more conservative position is that the role of government is to provide only true public goods, those goods which the private sector is unable to produce because no one can be excluded from using them or because users collectively consume the benefits. Goods and services should be produced by the private sector whenever possible; if not by companies for profit, then by non-profits and charitable institutions. It is a personal and family responsibility to care for family members in a nuclear family setting and more social programs encourage personal irresponsibility. The more liberal position is that if government provides a good or service to meet citizens' needs, the society as a whole benefits as well. During the George W. Bush administration there were efforts to promote more federal funding to faith-based organizations in an attempt to shift some of the social spending back to the private sector (The White House 2001). With continuing concerns over the size of the federal budget and debt, and with the housing and financial crisis that came to a head in 2008, the problem is going to be a continuing one.

Employment

Future employment and jobs prospects are at the core of economic growth. In the U.S. economy, two-thirds of the spending in the economy is consumption spending by private individuals (Economic Report of the President 2007). The jobs they hold and the after-tax income they receive from the jobs have a direct impact on their ability to participate in the economy. According to the Bureau of Economic Analysis, the trend of fewer goods-producing jobs and more service-providing jobs is expected to continue. Projections are that by 2014, service-providing jobs will outnumber goods-producing jobs nearly five to one (Bureau of Economic Analysis 2006). Of the top ten fastest-growing industries, four of the top five were health-care related. Fastest job growth—with a fifty percent growth rate—is predicted in community care facilities for the elderly and home health care services. While most of the increase is attributed to the graying of the population, part is also due to children of aging parents remaining in the workforce and having little time available for caring for aging parents and often family members live too far away to provide elder care (ODJFS 2006).

Occupations in technology will continue to experience growth, even though some jobs such as writing code will grow more slowly due to U.S. companies' outsourcing such tasks to lower-cost countries such as China. In the ten-year period ending in 1992, China had doubled the share of its gross domestic product spent on research and development. American companies such as General Electric, Microsoft, Cisco, IBM, Intel, and Daimler-Chrysler have facilities there devoted to research and development primarily because of the ready availability of highly trained science and engineering professionals (Garten 1966).

According to Martin Bailey and Diana Farrell of the McKinsey Global Institute, fears over massive job losses due to offshoring are unfounded. Even though researchers predict that 3.4 million business-processing jobs for U.S. firms will be done overseas by 2015, it is only a small part of the 150 million jobs that will be available in the U.S. economy (Bailey 2004). As many of the service jobs in the economy are knowledge-based, telecommuting, flextime, and job sharing are becoming more commonplace.

Robert D. Atkinson, vice president and director of the Technology and New Economy Project at the Progressive Policy Institute in Washington, D.C., argues that the "New Economy" will bring higher productivity that will allow people to have a healthier balance between work and non-work activities, with more time for family life (Atkinson 2006). The trend in the United States has been for new mothers to return to work soon after the birth of a child. In 1965 only seventeen percent of mothers of one-year old children worked; by 2001, that percentage had risen to fifty-eight. One survey found that seventy-two percent of U S. parents would rather stay home to raise their children, but could not afford to have one parent stay home. According to Atkinson, the New Economy will build a more humane economy by creating a more humane workplace which allows more time for the rest of life (2006).

As is usually the case, for every positive perspective, there is one which is less positive. In a review of Louis Uchitelle's book *The Disposable American: Layoffs and Their Consequences,* columnist Bob Herbert of *The New York Times* compares American workers to "battered soldiers at the end of a lost war." According to Herbert, for a large number of Americans the workplace is a place of anxiety and fear where nothing is certain. He quotes Mr. Uchitelle about the emotional aftermath of a layoff: "There's a lot of mental health damage . . . layoffs are a national phenomenon, a societal problem—but the laid-off workers blame themselves" (2006). He also argues that even better education and retraining will not result in better jobs for workers who are laid off.

In the September 2, 2006, on-line edition of *The Wall Street Journal*, a report card on the state of U.S. workers revealed that while the overall economy had grown by nearly twelve percent since the recession of 2001, median incomes had declined by one and three-tenths percent for men and one and eight-tenths for women. One of the most troubling developments has been the reduction or loss of pensions. Traditional advocates for workers' rights and benefits—labor unions—have seen their membership decline to only twelve and one-half percent of the U.S. workforce, down from thirty percent in 1970. There were 171,000 workers on strike in seventeen different strikes in 2004, down from 322,000 involved in forty-five strikes in 1994. Some observers argue that with such diminished influence, unions have outlived their usefulness and are ineffective in an environment where an average of 2.9 million temporary workers are employed each day, up from 2 million in 1995 (*The Wall Street Journal Online* 2006).

As with any social science, economics applies systematic principles to build models that project possible outcomes, but the fact that it is a "social" science, involving people and differing values, means there are no accurately predictable outcomes. Unforeseen events, such as the attacks of 9/11 and the collapse of financial institutions, can impact the economy and render predictions even less reliable than weather forecasts.

An Economy in Crisis

An unforeseen event that occurred just as this manuscript was ready to go to press was the need for, and the passage of, the Economic Stabilization Act of 2008. This was emergency legislation that allowed the government to buy illiquid assets and nationalize them in order to free up credit in the banking industry. The credit crisis started with Federal National Mortgage Association (Fannie), and the Federal Home Loan Mortgage Corp. (Freddie). These two corporations were government-sponsored enterprises (GSEs). They were privately-owned, in order to raise capital, but congressionally-chartered and controlled. Freddie and Fannie were subject to special oversight and yet, over time, violated many of the traditional rules of finance. These GSEs were urged to expand the percentage of homeowners, by making ownership accessible to a more varied economic class. Freddie and Fannie served as a secondary market and supplied funds for over one-half of the mortgages in the United States. They accepted loans from mortgage originators, bundled the loans and resold them or kept the loans in inventory. The government allowed them to work with very low capital requirements because they supplied a higher proportion of their loans to poor and minority groups. This allowed the originators of the loans to be lax in their lending practices because they had ready purchasers for their mortgages—Fannie and Freddie—and so did not have to keep the risky loans in their own portfolios. When housing became oversupplied, prices dropped and Fannie and Freddie became insolvent. This in turn created great fear of a possible domino effect throughout the financial industry, Wall Street, and even global markets. This crisis is projected to take years to overcome, yet the passage of the Economic Stabilization Act is expected to offer some "stabilization" to this crisis.

Future Trends—2020

Since we are now part of a global economy, our economy is vitally linked to other major economies of the world. Certainly variables such as terrorism, natural disasters, and political upheavals can alter economic trends. This notwithstanding, in 2020 the U.S. is expected to remain the most important single country because of the economic power resulting from the size of its Gross Domestic Product (GDP), its military might, and persistent technological lead. Even though the U.S. is expected to benefit less than some other countries, 2020 will still acknowledge the U.S. as a powerful economic force (Palmer 2006).

By the year 2020, Asia, with the fast growth of China and India, will dramatically narrow yet not close the gap in wealth, power, and influence. According to *Foresight 2020,* a report from the Economist Intelligence Unit sponsored by Cisco Systems, China is set to become the world's second largest consumer market. The world economy is expected to grow at an annual rate of three and one-half percent, as it has for the past twenty-five years, becoming two-thirds bigger in 2020 than it is now. Even though the European Union (EU) will have expanded to more than thirty members, encompassing all the Balkan Countries and Turkey, India will be rivaling the bigger European markets by 2020 (Palmer 2006).

Latin America will maintain an average growth rate of about three percent, yet the quality of its human capital is expected to lag behind emerging markets in Asia and Eastern Europe. For instance, the average wage in Brazil is now double that in China. By 2020, Brazil's average wage is expected to be about thirty percent below China's (Palmer 2006). Yet this does not mean that Brazil is a minor player on the global stage. In his article, "The BRICS Dream," Jim O'Neill, with the Goldman Sachs Group, suggests that the top ten global economies will have a different profile in 2050. According to O'Neill, BRICS (an acronym he introduced in 2001 that stands for Brazil, Russia, India, and China) will emerge as major economies. China's economy is expected to surpass that of the United States around 2035 and the Indian economy will join China in passing the U.S. economy by 2050 (2006). Nevertheless, if trends continue, the U.S. will remain the global leader as a result of the size of its GDP well beyond 2020.

The Soul of Economics—A Summary

Economics, the production and distribution of goods and services, is only a part of what brings human well-being and fulfillment. Since we live in an ever-changing world, why would we think that economics is exempt from the evolution and deterioration that overtake the rest of the world and its human institutions? Early economic thinkers did not begin as economists as we understand the discipline today. Many of them were philosophers, ethicists, moralists, and even clergy. Adam Smith, regarded as the father of modern capitalism, believed that members of a society should have the material means to live without shame. Smith even warned that no society can flourish when the far greater part of its members are poor and miserable (Smith 1776).

It does seem that the thrust of economic concern today ignores consideration of a society's spiritual dimensions and welfare. Dr. Alfredo Sfeir-Younis, Senior Advisor to the Managing Directors Office of the World Bank, says the world community is now demanding a transformation, an updated economic policy that pursues eradication of poverty, enhancement of social justice, and promotion of human empowerment. He voices concern that global economic practices have marginalized nearly half the planet, as more than two billion people earn less than two dollars a day. Sfeir-Younis says: "Too much attention has been paid to the 'human doing', the 'human having' and the 'human knowing', and much less to the 'human being'. While 'having', 'doing' and 'knowing' are extremely important, the intrinsic value, direction and identity of human life, like solidarity, caring, and sharing, are given by, and are in the nature of, the 'being'" (Sfeir-Younis, 2).

If one's expanded goal of economics includes enrichment, then a spiritual component of economics enhances the creative energy. Herein one moves from abundance as a quantitative idea to abundance as a qualitative one. In this, one may measure the economy not only in terms of goods and services but also in terms of empowerment of the human potential for creativity, meaning and fulfillment. This is a worthy inclusion indeed.

References

Atkinson, Robert D. 2006. "Building a More-Humane Economy." The *Futurist*. May/June 2006; Bethesda, MD: The World Future Society.

Bailey, Martin and Diana Farrell. 2004 . *The Milliken Institute Review*. Fourth Quarter, 2004, Milliken Institute.

Bailey, Thomas. 1971. *The American Pageant: A History of the Republic*. Lexington, MA: D.C. Heath.

Bryson, Ken and Lynne M. Casper. 1999. "Co-resident Grandparents and Grandchildren." *Current Population Reports*. Washington DC: U.S. Census Bureau. http://www.census.gov/prod/99pubs/p23-198.pdf. (accessed February 28, 2007).

Bureau of the Census, *Statistical Abstract of the United States*. 2004. http://www.census.gov/prod/www/statistical-abstract-us.html. Table 1330. (accessed January 10, 2007).

Bureau of Economic Analysis, *National Income Accounts*. 2004. http://www.bea.doc.gov/bea/dn/nipaweb/SelectTable.asp?Selected=N. Table 5.1. (accessed January 20, 2007).

Bureau of Economic Analysis. 2006. *BEA News*. http://www.bea.doc.gov/bea/dn/nipaweb/SelectTable.asp?Selected=N. Table 6.4. (accessed January 10, 2007).

CIA—The World Factbook. https://www.cia.gov/cia/publications/factbook/geos/kn.html#Econ. (accessed on January 7, 2007).

Congressional Budget Office. 2007. *The Budget and Economic Outlook: Fiscal Years 2008 to 2017*. www.cbo.gov. (accessed January 30, 2007).

Divine, Robert A., T. H. Breen, George M. Fredrickson, R. Hal Williams, Ariela J. Gross, and H. W. Brands. 2005. *America: Past and Present*. New York: Pearson/Longman.

Economic Report of the President. 1975. Washington, DC: Government Printing Office. www.gpoaccess.gov/eop/. (accessed January 30, 2007).

Economic Report of the President. 2005. http://www.gpoaccess.gov/eop/. Tables B-78 and B-79. (accessed January 31, 2007).

Economic Report of the President. 2007. http://www.gpoaccess.gov/eop/. Chart 4-1. (accessed January 30, 2007).

Fields, Jason. 2003. "Children's Living Arrangements and Characteristics." *Current Population Reports*. U.S. Census Bureau. http://www.census.gov/prod/2003pubs/p20-547.pdf. (accessed January 30, 2007).

Garraty, John A. 1966. *The American Nation: A History of the United States*. New York: Harper & Row.

Garten, Jeffrey. 2005. "The High-Tech Threat from China." *Business Week,* January 31, 2005.

Herbert, Bob. 2006. "Laid Off and Left Out." *The New York Times*. May 25, 2006. www.nytimes.com. (accessed January 10, 2007).

Keynes, John Maynard. 1936. *The General Theory of Employment, Interest, and Money*. London: Macmillan.

National Centers for Disease Control, National Center for Health Statistics. 2002. *Number and Percent of Births by race and Hispanic Origin, 1940-2002*. 2002. http://www.cdc.gov/nchs/data/statab/natfinal2002.annvol1_17.pdf. Table 1-17. (accessed Feb. 28, 2007).

National Centers for Disease Control, National Center for Health Statistics. "Number and Percentage of Births to Unmarried Women, all ages and women under the age of 20 years." United States Final 2004 and Preliminary 2005. 2005. http://www.cdc.gov/nchs/data/hestat/prelimbirths05_tables.pdf#1. Table 3. (accessed Feb. 28, 2007).

ODJFS. 2006. *Ohio Job Outlook 2014: Executive Summary*.

O'Neill, James. 2006. "The BRICS Dream." *The Goldman Sachs Group*. May 2006. www.2.goldmansachs.com/insight/research/reports/report32.html and www.2.goldmansachs.com/brics/brics_intro_content. (accessed March 13, 2007).

Palmer, Andrew, ed. 2006. "Foresight 2020: Economic, industry and corporate trends." *Economist Intelligence Unit.* San Jose: Cisco Systems.

Sfeir-Younis, Alfredo. 2001. "Embracing Spiritual Economics." *Kosmos Journal.* May 1, 2001. www.kosmosjournal.org/kjo/backissue/s2001/spiritual-economics.shtml. (accessed February 20, 2007).

Smith, Adam. 1776, reprint 1937. *An Inquiry into the Nature and Causes of the Wealth of Nations.* New York: Random House.

The Wall Street Journal Online. 2006 . "Labor Day: A Report Card for American Workers." http://online.wsj.com/article_print/SB115715685422352433.html. (accessed Jan. 4, 2007).

The White House. 2001. "Faith-Based and Community Initiatives: Compassion in Action." http://www.whitehouse.gov/government/fbci/guidance/intro.html. (accessed February 28, 2007).

Tucker, Irvin. 2006. *Survey of Economics.* Mason, OH: Thompson South-Western.

Part Two

Other Social Institutions

> We stand at the precipice of a new era. The economic and technological forces that have been building for more than two decades are about to crest, and thereby change our personal and social lives even more profoundly than they have been affected so far. There is no turning back to the old jobs and the old securities, to the old families and the old communities. So where do we turn? We delight in the terrific deals of the new economy. We stand in awe of its technological prowess. We are dazzled by the instant opportunities it presents for vast wealth. Yet where is the moral anchor? To what do we attach ourselves, our loyalties and our passions? Where do our friends, families, and communities come in? In what do we invest our integrity? How, in the end, shall we measure success—our own, as well as our society's?
>
> (Reich 2001, 247-248)

Chapter 6

Health Care

Robert L. Menz, DMin

The Hippocratic Oath

I swear by Apollo the healer, by Aesculapius, by Health and all the powers of healing, and call to witness all the gods and goddesses that I may keep this Oath and Promise to the best of my ability and judgment.

I will pay the same respect to my master in the Science as to my parents and share my life with him and pay all my debts to him. I will regard his sons as my brothers and teach them the Science, if they desire to learn it, without fee or contract. I will hand on precepts, lectures and all other learning to my sons, to those of my master and to those pupils duly apprenticed and sworn, and to none other.

I will use my power to help the sick to the best of my ability and judgment; I will abstain from harming or wronging any man by it.

I will not give a fatal draught to anyone if I am asked, nor will I suggest any such thing. Neither will I give a woman means to procure an abortion.

I will be chaste and religious in my life and in my practice.

I will not cut, even for the stone, but I will leave such procedures to the practitioners of that craft.

> Whenever I go into a house, I will go to help the sick and never with the intention of doing harm or injury. I will not abuse my position to indulge in sexual contacts with the bodies of women or of men, whether they be freemen or slaves.
>
> Whatever I see or hear, professionally or privately, which ought not to be divulged, I will keep secret and tell no one.
>
> If, therefore, I observe this Oath and do not violate it, may I prosper both in my life and in my profession, earning good repute among all men for all time. If I transgress and forswear this Oath, may my lot be otherwise.
>
> (Hippocrates of Cos II 460BC - 370BC)

Historical Perspective

Historically speaking, the sick have caused a sort of fear among humankind. Those who did not conform were dangerous and threatening and consequently removed from interaction. In primitive societies, the sick were considered non-conformists who had fallen out of divine order. In the Judeo-Christian tradition, primal disobedience in the Garden of Eden leads not only to banishment but to perpetual punishment. Thus, the "Fall" as revealed in the book of Genesis explains how suffering, disease, and death were the result of disobedience and "original sin." The ancient Greeks mythologically explained the presence of suffering and pestilence with the fable of Pandora's Box.

From Greco-Roman antiquity onwards, and also among the great Asian civilizations, the medical profession steadily replaced legendary explanations of illness with ones that became progressively nature-based. Roy Porter, in *The Greatest Benefit to Mankind,* suggests that from Hippocrates in the fifth century BC, through to Galen in the second century AD, "Humoral medicine stressed the analogy between the four elements of external nature (fire, water, air, and earth) and the four humors of bodily fluids (blood, phlegm, choler or yellow bile, and black bile), whose balance is health" (Porter 1997, 9). The goal was to seek balance in life and nature and indeed, the role of medicine was to restore this balance when lost. Thus, an imbalance would produce illness resulting in body fluids concentrating in a particular body area. That is, too much fluid in the legs may produce gout, in the lungs a cough. The healer's job was to restore equilibrium.

Long before the development of modern medicine, prehistoric humans tested medicinal plants, magical treatments and ritualistic charms. The Chinese, who saw mankind as a mirror of the universe and infused with the vital energy of chi, honor this understanding yet today. Even though Asian medicine is considered "nontraditional therapy," it is interesting to note that this tradition of healing dates back at least 3000 years.

In *Health and Wellness: Illness among Americans*, Margo Harris suggests that the oldest medicine known to mankind is called *Ayurvedic Medicine*. Ayurveda, which means "science of life," has been practiced in India and Asia for more than 5000 years. Ayurveda considers the fundamentals of diet, hygiene, sleep, lifestyle, yoga, exercise, and good relationships as key components to health. Harris states, "With an emphasis on preventing disease and promoting wellness, its practitioners view emotional health and spiritual balance as vital for physical health and disease prevention" (Harris 2005, 149).

Even in more modern times, the practice of health care can be quite varied. The Native American medicine man's practice included the sweat lodge, rhythm of the drums and use of magic plants. Tibetan healers meditate the fate of their patients by becoming divine and manipulating time. In the Andes of Peru, the one who survived being struck by lightning became the interpreter of ailments through the reading of coco leaves (Smolan, Moffitt, Naythons 1990, 12). Are these poster examples for the scientific method? I think not. However, there may be value in acknowledging the community aspect of collective and preventative wellbeing. There is also a sacred component to health. "Sickness is disruption, imbalance, and the manifestation of malevolent forces in the flesh. Health is a state of balance, of harmony, and for most societies, it is something holy" (1990, 11-12). So when we are well, we are whole.

The word health comes from the Old English word "hale" from which we get the words whole, and even holy. Medicine comes from the Latin *medicina*, which is derived from an Indo-European root that also gives us the words *remedy* and *meditate*. The word *disease* comes from the Old French word meaning "lack of ease" (Weil 1998). Being uneasy or imbalanced or lacking ease is certainly a fitting description for disease. Whereas *health* describes the given state of the human condition, *wellness* involves an active pursuit of wholeness which reflects positive conditions in all aspects of a person (i.e., mind, body, spirit, job, family, and environment). Often in the West, however, the body is not so much seen as a pulsating form of homeostatic energy as it is a complex machine in good or bad repair.

Early hospitals were almshouses or hostel-type religious foundations that extended acts of compassion. However, in the late seventeenth and early eighteenth centuries, as anesthetics and antiseptics moved beyond opium and alcohol, hospitals began a transition from charitable shelters for the sick poor to medical institutions addressing the gamut of needs for the public at large.

Doctors not only cared for the sick—sometimes hours at bedside—they also taught patients and family members how to care for themselves naturally. As a matter of fact, the word *doctor* comes from the Latin for *teacher*. This concept seems to have been abandoned. The current model of a doctors' office certainly does not lend itself to "teaching." Indeed, many times a nurse will check a patient's blood pressure and not even share the reading with the patient. Doctors sometimes teach in hospitals during *rounds*, but the patients are certainly not the students.

Current State of Medical Care

In modern Western urban societies, *Allopathic Medicine* is the only theory of medicine taken seriously. Allopathy is the dominant system of "medicine" in the world. When it comes to acute medical conditions, allopathy is considered to be the treatment of choice. However, many chronic conditions that respond to the emotional and spiritual aspects of a person may be ignored by allopathic—conventional—medicine.

Today's distinction between physician and surgeon, with each having many new subdivisions and specialties, was not present among the ancient Greeks and actually did not start appearing until the Middle Ages. Today, the dissection of the medical profession may encourage doctors to utilize their technical abilities to rely more on chemicals, laboratories, and computerized tests than to cultivate their powers of observation, intuition, and indeed, of healing.

Andrew Weil is a medical doctor world-renowned for this holistic approach to health. In *Health and Healing* Weil states:

> I would look elsewhere than conventional medicine for help if I contracted a severe viral disease like hepatitis or polio, or a metabolic disease like type 2 diabetes. I would not seek allopathic treatment for cancer, except for a few varieties, or for such chronic ailments as

> arthritis, asthma, multiple sclerosis, or for many other chronic diseases of the digestive, circulatory, musculoskeletal, and nervous systems. Although allopaths give lip service to the concept of preventive medicine, for practical purposes they are unable to prevent most of the diseases that disable and kill people today. (1998, 83)

In *The Natural Mind*, Weil advocates using some of the ancient wisdom of non-material thinking characteristic of Eastern traditions, along with linear and concrete thinking of the West. Perhaps the complementary nature of the two approaches will eventually move toward synthesis. Weil says, ". . . the goal I aim for is a total synthesis of Eastern and Western approaches to the mind. Consciousness is real. So is the physical nervous system. They are two expressions of the same underlying unitive phenomenon, and we must pay attention to both of them if we are to experience that phenomenon as a single reality" (Weil 2004, 152).

We must admit that no one source seems to have all the answers. Not so long ago we were told that Pluto was the last object orbiting the sun, that atoms were the smallest particles, and perhaps there were four dimensions. Now science considers that the eleventh dimension may be generating a parallel universe. Likewise, our pursuits in chemical medicine may someday be considered as limiting to health as was "blood letting" for President George Washington.

Alternative forms of medical care seem to be taking the place of the folk medicine of yesteryear. The discipline of homeopathy, which uses small doses of drugs and seeks to engage the body's own natural defenses, has great opposition from the medical establishment and particularly the powerful drug companies. Osteopathy and chiropractic, which manipulate bones and joints in an effort to regain balance and alignment, are not as great a threat to the medical establishment as homeopathy, and consequently, are regarded with greater esteem. The Chinese art of acupuncture, a healing technique in which fine needles, or *lances*, are inserted into key spots on the skin to keep the body's life energy in harmony, is used in the west as an alternative treatment (sometimes with great success); in China acupuncture is widely used and respected. More complementary than alternative are the mind, body, and spirit exercises such as stress reduction techniques, yoga, meditation or prayer, and hypnosis.

Acknowledging great medical breakthroughs like bacteriology and modern technological imaging, one must still concede that some of the greatest contributions to the advancement to human health have come from outside the discipline of medicine. Such things as screens on windows, clean water, proper hygiene, proper nutrition, seat belts, and helmets have saved countless people from disease, disability, or death. Awareness of environmental hazards has saved limbs and avoided cancers. Being aware of what contributes to a healthy lifestyle and being alert to hazardous situations promotes personal health.

As I wrote in a previous book, one may think of health intervention on four levels.

> The first level is *medical intervention*. When the body is broken, get it fixed! This level, of course, is the traditional model for intervention. This level will always be available to us, yet historically our society has too often taken health for granted until it is lost.
>
> Another level is *structural prevention* in which attention is given to changing the society or environment within which people work and live. Examples include speed limits on the roadways, seat belts in automobiles, and smoke detectors in homes. Over the past decade, significant attention has been given to such issues as ergonomic awareness and laws regulating drinking alcohol and driving. Incorporating a healthy mind-set within the very structure of organizations will build a healthy foundation for our children and their children.
>
> The third level, *pro-wellness*, fosters a social climate of well-being. Pro-wellness is primarily conceptual and involves advocating healthy choices and avoiding unhealthy practices such as a company that incorporates wellness practices into all levels of the organization. After all, the best way to stay well is to avoid sickness and injury.
>
> Finally, *behavioral prevention* involves claiming ultimate authority and responsibility over personal wellness. These individuals have become convinced that they alone are ultimately responsible for their wellbeing. Knowledge is power, and understanding is healthy. Most employers have utilized the first three interventions with degrees of success. Accomplishing behavioral prevention reaps substantial benefits. (Menz 2003, 12-13)

The deadly diseases dreaded at the turn of the twentieth century—tuberculosis, tetanus, rheumatic fever, meningitis, pneumonia, polio, and syphilis—no longer cause such fear as we move into the twenty-first.

Today surgery may be accomplished through a tiny incision only after the surgeon has already seen the surgery site through CAT scans, PETT scans, bone scans, MRIs, lasers, X-rays, or ultrasound. Medicine can now minimize, if not eliminate, pain. We are immunized against diseases that used to terrorize. Where would we be without antibiotics, steroids, psychotropics, and immunosuppressants? Yet, although we have conquered many maladies, suffering remains and death, though delayed, awaits.

In *The Greatest Benefit to Mankind*, Roy Porter suggests that the future of medicine has become muddled. What is its new mandate? Where will certain pursuits lead? Porter asks, "Is its charge to make people lead healthy lives? Or is it but a service industry, on tap to fulfill whatever fantasies its clients may frame for their bodies, be they cosmetic surgery and designer bodies or the longing of post-menopausal women to have babies?" (Porter 1997, 717).

People tend to share and promote their own paradigm, to give what they have. In health care, for instance, a nutritionist will suggest certain nutrients due to a perceived deficiency, a physical therapist will talk about specific exercises to target certain muscles, a psychiatrist will likely prescribe an antidepressant medication for depression, a psychologist will see psychotherapy as the likely remedy for depression, a chiropractor will discuss the importance of alignment for back pain, and a surgeon will prefer surgery to other specialties. If a treatment is not understood, like acupuncture, reiki, or energy flow, it may be ridiculed. We all tend to be afraid of what we don't understand, but some even seek to make the treatment or caregiver look shamefully ridiculous—if not stupid. Although all forms of health care used need to be proven safe and effective, some interventions endorsed by the American Medical Association (AMA) have occasionally been found not to be safe and/or effective. One wonders whether there is a discipline anywhere that seeks to coordinate the various forms of health care intervention. If the word *doctor* means *teacher*, one asks, "Where then is the teacher of wholeness?"

Noteworthy also is the fact that medical intervention has evolved over time. Opinions over certain ailments and cares change through time. I know a woman who once suffered from endometriosis and a mass grew on one of her ovaries. The surgeon said, "Let's do a total hysterectomy. After all, your uterus has completed its function of carrying babies," conclusively suggesting that the uterus has no other value. Coming to the awareness that an organ or gland has purpose beyond the obvious one is similar to the issue of "object permanence" where an infant learns that an object exists even if it cannot be seen. Indeed glands and organs may have purpose well beyond the obvious. The woman chose to have the mass removed, yet refused to have her healthy uterus removed. She recovered well and avoided the further complications of a total hysterectomy.

Few would argue that there have been times when body organs were removed "in order to be safe" much more often than they are today. Who knows how many tonsillectomies, hysterectomies, radical mastectomies, and shaved arthritic knees were performed before today's standards began to suggest: "Not so fast with the knife!" After all, the uterus is more than an incubator.

Forty or fifty years ago, most people sought medical care for acute medical conditions such as broken bones, lacerations, infections, and organ failure. Now the public rushes to medical facilities for chronic medical conditions like diabetes, cancer, high blood pressure, and high cholesterol. Many of the common and mostly unavoidable ailments are now becoming "medicalized." Mom or grandma used to fix us a cup of chicken soup and we would be back to normal in a week. Now we may take antibiotics for very common illnesses (sometimes for viruses) and get back to normal in a week. This is not to minimize the need and benefit of the medical community, but we are not helping the doctors or ourselves if we over-medicalize routine ailments. Where is the room for personal growth and character building if, for instance, "road rage" is viewed as a medical disorder? We may also need to clarify that to be quiet is not necessarily to be depressed; to be rowdy is not necessarily attention-deficit hyperactivity disorder; and to be bald is not abnormal, although such promotion of vanity may be.

Larry Stepnick reported on a conference convened by the National Institute for Health Care Management in 2002 where prominent stakeholders, employers, health plans, federal and state governments, and consumers gathered to discuss accelerating the adoption of preventative health services. Even though there are lingering doubts about the clinical benefits and cost-effectiveness of many preventative care services, growing data are revealing measurable return on investment (ROI). For instance, selected preventive services, such as smoking cessation and disease screenings, yield an ROI of from two to four dollars for every dollar invested—a return of from two- to four-hundred percent (Stepnick 2003).

Currently, Medicare helps pay for most drugs for medication but offers help for few forms of preventative care, wellness education, or better nursing staff ratios in hospitals and nursing homes. New tests, medications, and technological tools may be of marginal or no improvement of the collective health of our nation; yet it sometimes seems we are sacrificing fundamentals in pursuit of the latest medical pill or procedure. Deyo and Patrick report: "The United States spends more than any other country on health care, but has shorter longevity, worse public health statistics, and worse health care inflation than most developed nations" (2005, 4). We in the United States are particularly vulnerable to the notion that if we just spend more and lean into the cutting edge of technology, we will prevent illness. We are not being scientific or "critical" in our thinking if we continue down the recent path of the medical enterprise like Ponce DeLeon pursuing the fountain of youth.

Further need exists in the field of health information technology (HIT). Electronically connecting health records and technology to individual health care may yield only a modest ROI in the short term. Yet the long-term need is unquestioned. Even though the recent implementation of the Health Insurance Portability and Accountability Act (HIPAA) is not yet showing a ROI, it lays a foundation vital for standardized business transactions in health care.

Leading the way and setting an example in electronic medical records is the VA Medical system in the United States. As a veteran who has accessed this medical records service, this writer appreciates the efficacy of this system. In the past few years the VA health care system has been rated exceptionally high (even by patients); and there is an emphasis on prevention, yet, the greatest advantage for the VA is their electronic medical records. The VA has the largest and one of the most modern systems in the world. When a VA patient visits any VA facility in the country, the records are accessible there. All the information available on any VA patient is at the fingertips of the health care provider, including medical history, current condition, any X-rays, lab results, or tests that are planned. The electronic records also allow the VA to track its own performance—to quickly learn what works and what does not. Its health system may indeed be a model for health care for our entire country. For those who have been referred to a series of medical specialists or have gone through layers of medical personnel answering the same questions over and over again when being admitted to a hospital, this need is evident. The seemingly disconnected and inefficient nature of this process can be very frustrating.

Problematic Issues

Only a small portion of the total health care attention and dollars are directed toward preventative medicine. In fact, tests for chemical profiles, mammograms, colorectal screening, cervical cancer screening, and PSA tests for prostate cancer, are really pursuits of early detection rather than prevention of disease. Doctors find themselves in a culture of disease treatment more than health care. Most doctors, if following their namesake, *teacher*, are teaching how to *get* well rather than how to *stay* well. Were medical doctors to dedicate as much time and attention to prevention as dentists do, we would have a healthier society. Indeed, our bodies contain many of the remedies for what ails humanity if we could activate these defenses. Yet it seems to be the way of humanity to seek external solutions by assuming that medical cure must be externally and pharmaceutically produced. For the last two or three decades, it seems that medicine has become the hinge pin for wellness and the pharmaceutical companies have become the proverbial tail wagging the dog of health care.

If the medical community could direct more attention to the oxidation process that exists on our planet, perhaps great strides could be made toward the prevention of disease. It is the same corrosive oxidation process that causes iron to rust, the ozonosphere to be damaged, or a freshly peeled apple to turn brown that is involved in the oxidizing forces leading to such degenerative diseases as heart and artery disease, cancer, autoimmune diseases, and arthritis.

Better than the question, "What drug out there may help me?" is the question, "How may I activate my healing response?" Yes, we spend a lot of money on our health care system but are we getting sufficient "bang for our buck"? Actually, no. "The health-care system we claim is the best in the world is actually near the worst when you look at how long Americans live—or don't live" (Strand 2002, 11).

We live in a world where we can have movies on demand, food in an instant, fly at nearly the speed of sound, and pop a pill for a sundry ailments. Quick fixes are available for pain, which only mask the problem, and for lowering cholesterol, while in some cases implying that you can go ahead with your unhealthy diet. It is surprising to learn that cholesterol is not the sole culprit behind heart disease; inflammation of blood vessels, however, is of major concern. Of course the marketing of cholesterol-lowering medicine is not wrong. Nevertheless, the lack of education about the inflammation of blood vessels may be. It is interesting that more than half of heart attack patients in the United States have normal cholesterol levels (Strand, 2002). In addition, by reducing free radicals and by taking antioxidants, one can lower the inflammation of the blood vessels, thus lowering the risk of heart disease and heart attacks. Nothing expensive or high tech here though, so this information is not greatly promoted. Thirty years ago cholesterol levels of 300 were considered normal. Now it has been lowered to below 200. Interestingly, the packets of information that accompany all the cholesterol-lowering medicines that this writer has seen state that the medicine does not claim to prevent heart disease or heart attacks. Given all the variables involved in the human system, one can appreciate the need for this statement. Nonetheless, one could benefit from knowing the statistical rates of improvement that cholesterol-lowering medicine does have on cardiovascular health.

Indeed we want our high density lipoproteins (HDL) to be high and our low density lipoproteins (LDL) to be low, but inflammation in our bodies, and especially in our blood vessels, is an important factor as well. In *What Your Doctor Doesn't Know about Nutritional Medicine*, Dr. Ray Strand quotes a study that reveals that patients with the highest levels of antioxidants in their bodies had the lowest occurrence of coronary artery disease (2002). Consequently, many cardiologists are busy treating patients who are at the end of the inflammatory process, and then do not have sufficient time to *teach* other patients who are moving toward greater cardiovascular damage, how to minimize inflammation. By the way, Dr. Strand's book not only offers an excellent overview of antioxidants like vitamins E and C, and CoQ10, but also suggests some new tests to detect heart disease like the highly sensitive C-Reactive Protein (hs CRP), homocysteine blood levels, and heart calcification levels. It would be good to hear more about these tests from the medical community.

In addition, a growing number of studies are revealing the benefits of natural remedies and over-the-counter medicine and supplements for health enhancements. A recent study initiated by the National Institutes of Health (NIH) revealed that the popular joint health compound Glucosamine and Chondrotin has significant potential to support joint health and comfort. The trial analyzed thousands of participants over the age of forty and confirmed what was already revealed in earlier studies: Glucosamine and Chondrotin aids in making stronger and healthier joints (Bradzey, 2006).

One may question whether the pharmaceutical companies are in the business of promoting health or in the business of promoting their product. If the physician only relies on the drug sales representatives and the drug companies' curricula for information, there will be a hole in the physician's understanding of integrative wellness. Lest we forget, the difference between prescription drugs and poison is *amount*. The Greek word *pharmakos*, from which we get the word *pharmacy*, meant both remedy and poison. Potent pharmaceuticals walk a fine line between "cure" and "kill." This is important because potent medicines that were once only given in hospitals now are prescribed for in-home use where one may carelessly mix these medications with alcohol and/or other drugs. Because of this, it is reported in a recent AARP Bulletin, that accidental deaths from improper use of medication rose over 700 percent from 1983 to 2004 (Barry, 2008).

In *Overdose*, physician Jay Cohen reports, "After more than a decade of research conducted without any influence, I found that the drug companies dominate the entire process of medication therapy—from early research to ultimate usage—as few other industries control their products today" (2001, 9). Cohen further suggests that some of the medications that have been removed from the market since 1997, including Rezulin, Lotronex, Propulsid, Sedux, Pondimin, Duract, Seldane, Hismanal, Posicor, and Raxar, have resulted in thousands of deaths. Also, life-threatening side effects from medications and multiple emergency surgeries have reaped the pharmaceutical companies billions of dollars in revenue. Alarming examples are the drugs encainide and flecaindie. An estimated 50,000 people died from taking these medications that were intended to prevent cardiac arrest (Moore 1995). In addition, it is becoming commonplace for monies to go for more costly and less effective drugs. Even some of the new blood pressure medications have not been proven superior to legacy diuretic therapy for high blood pressure (Moore 1995).

In 2000, the anti-arthritis drug Vioxx was the most heavily advertised drug to consumers. The multinational drug conglomerate Merck spent more on direct-to-consumer advertising for Vioxx than was spent by PepsiCo, Budweiser Beer, and Dell Computer in support of each of their sales (NIH 2000). This was in addition to the way Merck marketed the drug to physicians who then prescribed it to more than 20 million people in the United States, and 80 million worldwide (Badzey 2006). Five years later, Merck was compelled to withdraw the painkiller from the market after it was linked to deadly cardiovascular problems. The question of how much Merck knew about these potential problems will likely be settled in the courts. Merck will likely face billion of dollars in state and federal lawsuits and the prospect of litigation over several years (Johnson 2007). It may not be until lawsuits are levied against misleading drug advertising, as they were against the tobacco industry, that full truth and disclosure will result. As it stands now, some of the claims being made in pharmaceutical commercials are beginning to take on the tone of the "Step Right Up" carnival-like medicine shows of a century ago.

Also, higher doses produce higher efficacy rates, which are great for marketing, even though these higher doses produce the above-mentioned side effects even as alcohol can. Some high doses of certain medications can even impair one's functioning. It stands to reason that too much of a substance can be harmful. If a glass of wine per day won't hurt you, a case of wine per day might! "Super sizing" fast food may be harmful, bigger doses of medicine may be worse. Furthermore, drug companies tailor the dosage of their drugs for the convenience of the doctor's busy schedule and the research and marketing needs of the pharmaceutical company.

Most alarming is that drug companies influence drug research and the information provided to doctors and consumers. Cohen reports that drug companies actually exaggerate data that they want to promote, suppress information that is not favorable, manipulate measurements to produce the best profile, and generally mislead by slanting or spinning information to advertise their product (Cohen 2001). This is the reason that prescription drugs are the fastest growing portion of health care costs, rising many times above the average rate of inflation (Strand, 2002).

Some of the medical "breakthroughs" are beneficial even if the "cost-benefit" ratio is slight. However, some procedures and medication are very costly yet may have significant side effects and still not shift one's health in the desired direction. In fact, many of the drugs can be (and are) slightly altered chemically and sold as the newest, and implied best, drug in their class. These so called "me too" drugs are then marketed to create new demand while charging even higher prices and crowding out the cheaper alternatives. Indeed some of the alternatives, even generics, have been proven to be more effective and safer. Some of the marketing tactics seek to persuade the public to demand products that they don't need.

The U.S. consumer cannot avoid the growing advertising campaigns of the large pharmaceutical companies. Their marketing in journals, magazines, and on radio and television is overwhelming. On a given evening, one may notice more pharmaceutical commercials on television than car, beer, and soft drink commercials. At a television commercial break one may hear three separate prescription drug testimonials. The National Institute for Health Care Management (NIHCM) reported that the pharmaceutical companies spent $2.5 billion in 2000— up from $791 million in 1996—on direct-to-consumer promotion, with fifty-seven percent of that spent in television advertising. This, however, is only the beginning of their promotional spending. After all, the primary customers for the prescription drug companies are the doctors. For the doctors, these companies spent nearly $12 billion on doctors' office detailing and free drug samples. These companies treat doctors (and the doctors' staffs) to free meals, vacations, wine tasting, and various entertainment venues (NIHCM 2000). Recently, change appears to be in the offing, because doctors and the public alike are raising conflict of interest issues with these practices. It is likely that in the near future state and federal laws may address this problem and regulate limits on gifts. But beyond these obvious influences, the drug companies manipulate both doctors and consumers by controlling all the published information about the drugs themselves. Package inserts, advertising and the information in the universal medications guide, the *Physician's Desk Reference* (PDR 2008), all come from the drug companies. Since these companies want to sell their product, there is a potential risk that this information could be biased or incomplete.

In *Overdose,* Jay Cohen offers one solution to the problem, as well as the issue of drug effectiveness and safety: creation of a monitoring agency independent of the Food and Drug Administration (FDA). Just as the National Transportation Safety Board is separate from the Federal Aviation Administration, a National Medication Safety Board could help protect the drug industry, the medical arena, as well as the public (Cohen 2001).

In *Hope or Hype* Deyo and Patrick, take the idea even further, proposing required, comprehensive labeling for all medications. This labeling process would state clearly such things as what a drug is for; who should consider using it; how long the drug has been approved; how it acquired approval; results of research; the number of people tested; and a comparison with placebo results. If a drug company is making the claim that "this drug will reduce your risk of stroke by fifty percent," the consumer could see whether that meant from sixty to thirty percent or from one percent to a half one percent (2005, 281-282). Today we are vulnerable to the "spin" driven by the sales motivation.

The FDA approves drugs and medical devices and seeks to assure the safety of food and cosmetics. The FDA monitors over $1 trillion worth of products representing nearly a fourth of all consumer spending (Deyo and Patrick 2005). From a medical perspective, the FDA is to protect the public from ineffective or harmful products. For a new drug to be approved by the FDA, it must be effective, fairly safe, but not necessarily better than an existing drug. In some cases "effective" tells a consumer little or nothing about the product. For instance, "A drug that lowers cholesterol, lowers high blood pressure, or improves heart rhythm can be approved without knowing if it improves life expectancy or quality of life" (Deyo and Patrick 2005, 154).

It is commonly reported that the top three leading causes of death are heart disease, cancer and cerebrovascular disease. Frequency of other causes of death varies greatly, depending on what sex or race is being considered. Overall, however, the remainder of the top ten causes of death in the United States are chronic lower respiratory disease, unintended injuries, diabetes mellitus, pneumonia and influenza, Alzheimer's disease, nephritis, and septicemia (Wexler 2003). Less-publicized information includes extensive studies showing that even properly prescribed and administered medications cause adverse reactions leading to more than 100,000 deaths in the U.S. each year (Bates 1995; Lazarous, Pomeranz, Corey 1998; Cohen 2001; Strand 2002). These data make medication reactions the fourth leading cause of death in the United States, outnumbering automobile accidents, AIDS, alcohol and illicit drug use, infectious disease, diabetes, or murder. Furthermore, there are over 2 million severe medication reactions annually in hospitals in the United States (Cohen 2001). Consequently, it is confusing when the medical community discourages patients from taking supplements on the basis that supplements could be dangerous to their health.

O, The Oprah Magazine reports William Prey, MD as saying physicians have become too comfortable with prescribing antidepressants. That is, "They're prescribing them to treat conditions that would have previously merited a stiff upper lip" (Ramin 2006, 255). Fifty years ago only severely depressed people were prescribed antidepressants. Now antidepressants are being prescribed for an ever-expanding variety of circumstances including shyness, eating disorders, premature ejaculation, sexual addictions and smoking. Joseph Glennullen, MD, clinical instructor in psychiatry at Harvard Medical School, suggests that the idea of medication was to improve the patient's emotional condition to the point of being able to take charge of his or her life, often with the help of psychotherapy. Albeit, "somehow that thought was lost, replaced with the idea that the drugs alone can do the job" (Ramin 2006, 284).

This concern is also shared by William Glasser, MD, a world-renowned psychiatrist, who developed *Reality Therapy* and *Choice Theory,* both well-known in the fields of psychiatry and psychotherapy. In his book, *Warning:Psychiatry Can Be Hazardous to Your Health,* Glasser reveals the dark side of biological psychiatry by providing the reader with a large body of scientific research that ". . . points out these drugs are nowhere nearly as effective as is claimed by the companies that make them" (Glasser 2003, 3). How can the homeostatic ebb and flow of brain chemicals find balance when they are hijacked by foreign drugs?

In addition to the idea of comprehensive labeling of medicine and a National Medicine Safety Board, another process that will help the excessive prescription problem is the promising science of "Evidence-based Medicine." This new approach will change medical practice in the years ahead. Evidence-based medicine challenges statements like, "I've used this medication for months now and my patients seem to do very well with it." This statement reflects the art of medicine but not necessarily the science. For instance, could apparent improvement be due to other factors? Had the illness run its course? Could there be other healing actions initiated by the patient? Could there be a placebo effect working? Evidence-based medicine moves beyond experience and expertise (though both are necessary and highly respected) to also employ the scientific method to assure reliable and valid treatment (Eisenberg 2001).

Looming Health Concerns

From the time of the early cities in Mesopotamia and Egypt, humans were menaced by diseases resulting from close proximity to animals. Unbeknown to humanity until the last two centuries, pigs and ducks were giving humans influenzas; horses, the rhinoviruses (the common cold); and dogs (through a canine distemper called rinderpest), the measles. In addition, innumerable people have died from animal bacteria like salmonella and animal feces that spread polio, cholera, typhoid, viral hepatitis, whooping cough and diphtheria (Porter 1997). Today more than 200 million people live in countries where there is life expectancy of less than forty-five years. In some of the poorest countries of the world, one in five children fails to reach his or her fifth birthday. The primary reason is infectious disease related to the environment (WHO 2004). Yet the highly developed cultures of the world offer a stark contrast. For instance, a girl born in Japan in the first decade of the twenty-first century will have a fifty percent chance of seeing the twenty-second century (Tischler 2007).

Influenza

Looming on the horizon is the threat of a deadly contagion unlike any of us can fathom. The last great pandemic was the 1918 influenza virus, which probably emerged in the United States, and ultimately killed an estimated 50 million people worldwide. It is reported that "Influenza killed more people in a year than the Black Death of the Middle Ages killed in a century; it killed more people in twenty-four weeks than AIDS has killed in tweny-four years" (Barry 2004, 5).

Of current concern to the Centers for Disease Control and Prevention (CDC), is the so-called bird flu, *avian influenza A (H5N1)*. Of the few avian influenza viruses that have crossed the species barrier and infected humans, this one, with outbreaks in Asia and Europe, has caused death in half of those infected (CDC 2006). Even though no one can predict when a pandemic might occur, experts from around the world are watching this virus very closely. The ultimate victor in the battle between *homo sapiens* and microbes is yet to be revealed.

AIDS

Of continuing concern is the Acquired Immunodeficiency Syndrome (AIDS), first reported to the CDC in summer 1981. Since that time, few medical problems have been investigated with such intensity. The CDC reported an eight percent decline in deaths from 2000 to 2004. Indeed, at the end of 2004, an estimated 415,193 persons in the United State were *living* with AIDS and up to 2004, a total of 529,113 had died from AIDS. World wide, approximately 40 million people have contracted the virus and slightly over half of them have died (CDC 2006a).

Equally significant is what has been learned regarding society's reaction to the deadly disease, primarily in the United States. From the beginning of the AIDS problem, the disease was labeled in moralistic terms (Menz 1987). As if the virus knew it was going after the "immoral" of society or that God was punishing the homosexual or drug user, society assumed a posture of repulsion. Many times through history there have been diseases that scared the masses. From leprosy in Mesopotamia or the Holy Land in Biblical times, to the bubonic plague among the Philistines, to the Black Death in Europe, to AIDS today, humanity has examples of condemnation for the sufferer. Nevertheless, there have also been voices to challenge this kind of thinking. The book of Job clearly challenges this theological point of view which says that suffering is a direct result of one's sin. Although the Hebrew Bible Book, *Job*, was not unknown in the time of Jesus, it apparently had little influence. The old theory of the cause-and-effect relationship between sin and sickness still prevailed. It was as though Jesus' voice stood alone when he rejected the simple formula that suffering was a sign of sin. If we think and act with compassion, the good of our humanity surfaces. When we insist on seeing things from a single point of view without thinking them through, or without following a line of thought to its natural consequences, we can often draw inappropriate conclusions. It is easier for the person not suffering to assume a relationship between sin and suffering than it is for a suffering "Job" to accept that claim. When we fail to "unpack" a concept—fail to set it forth in the open and consider the logical order—we simply cannot see its implications. This is what enables some of us at times to hold completely inconsistent, or even contradictory, views.

Lest we forget, assistance, compassion, and love are very powerful forces for well-being. Our society and its social institutions can take great strides by removing both the stigma from illness and greedy conglomerates from health care. The more society unites for that which is right, the better it manifests the potential goodness of mankind.

Obesity

Another major health issue, particularly in the United States is the increasing problem of obesity. An April 2005 report generated by the National Institute for Health Care Management states that two-thirds of the United States adult population and fifteen percent of its children and adolescents are overweight. Among adults, being obese is equivalent to aging twenty years. The total direct and indirect costs attributed to obesity may be as high as $117 billion annually and account for more than twenty-five percent of the increase in health care costs between 1987 and 2001 (2005).

The campaign against tobacco use has caused a reduction in tobacco smoking. However, the problem of obesity among adults has continued to climb in recent years. In fact, a large number of overweight children are developing obesity-related medical conditions like diabetes, high cholesterol, and high blood pressure, which were previously considered rare (Harris 2005).

The CDC has been tracking obesity for over twenty years and is charting a dramatic increase in obesity in the United States. In fact, the states of Texas, Louisiana, Mississippi, Alabama, Arkansas, Tennessee, Kentucky, West Virginia, and Michigan are faring the worst (2004a). Pauline Jelinek, a writer for the Associated Press reported that the U.S. Army is rejecting recruits for being overweight "couch potatoes" (Jelinek 2006).

The best strategy for solving the obesity problem is to unite the concerns and efforts of individuals, families, employers, health plans, schools, and governments. Perhaps through such collaboration, this health issue can be corrected.

Tobacco

One of the leading contributors to morbidity and mortality continues to be tobacco consumption. In 2004, the office of Surgeon General Richard Carmona issued that organization's Twenty-eighth Report on Negative Health Effects of Smoking (CDC 2004b). During the more than forty years since the first such report in 1964, great strides have been made toward educating the public about the harm of tobacco. This Twenty-eighth Report revealed that the per capita annual adult consumption in the United States had declined from 4,345 cigarettes in 1963 to 1,979 cigarettes in 2002. In fact, in that year, there were more former smokers than current smokers in the United States.

Fifty years ago smoking was a "rite of passage" for most youths and was permitted almost everywhere—including hospitals. Smoking among high school seniors has declined from more than thirty six percent in 1997 to less than twenty-five percent in 2003. Now that second-hand smoke is recognized as harmful, many workplaces and public spaces ban smoking. Some states like California have had aggressive tobacco control policies since 1989. The state of Ohio initiated anti-smoking laws for public spaces in 2007. The nation of Spain outlawed smoking at work, effective January 2006 (Journal of Employee Assistance 2006).

Yet the Surgeon General's report suggests that more effort is needed. Smoking is still the leading preventable cause of disease and death in the United States, leading to more than 440,000 premature deaths each year. Furthermore, current estimates of attributable economic costs in the United States are over $157 billion yearly.

One hundred years ago, lung cancer was a rare disease; now it is the leading cause of cancer deaths in both men and women. In the forty-plus years that tobacco has been a concern of the Surgeon General's Office, smoking has been identified as the cause of ten different kinds of cancer, four distinct types of cardiovascular diseases, six respiratory diseases, four negative reproductive effects, and five other negative effects (such as cataracts and low bone density). The report concludes that smoking harms nearly every organ of the body and reduces general health even in the absence of an identified disease.

The good news is that quitting smoking has immediate as well as long-term benefits no matter how long one has smoked. In addition, there are good "stop-smoking" techniques, aids, and even medications that help those motivated to quit. Indeed, as we move toward the year 2020, it is likely that people with unhealthy lifestyles will be expected to pick up a greater portion of the health care tab by being charged larger health insurance premiums.

Future Trends—2020

The financial burden of health care is becoming progressively unhealthy. In a summary of *Medicare and Medicaid: Projected Expenditures*, authors Earl Dirk Hoffman, Jr., Barbara S.Klees, and Catherine A. Curtis project that national health expenditures will reach $3.6 trillion in 2014, up from $1.7 trillion in 2003. This spending will reflect 18.7 percent of the gross domestic product (GDP) in 2014, up from 15.3 percent in 2003 (Hoffman, Klees, and Curtis 2005). If this trend continues, by the year 2020, we could be spending one fourth of our GDP on health care.

The rising cost of health care has given rise to a variety of health management organizations and other cost containment initiatives. The introduction of tax-advantaged portable Health Savings Accounts (HSAs) in the context of the Medicare Modernization Act of 2003 is a recent option in the field of consumer-driven health care. HSAs allow the consumer, through his or her employer, to set aside tax-free money to be applied to health care costs. Early 2005 enrollment numbers indicate that these savings accounts doubled within one year (Ehrbeck and Packard 2005). Will the HSAs last? It actually looks promising.

If we do not address some of the problems in health care, the crisis will worsen. The facts are that the number of people without insurance is growing; the burden of employer-sponsored insurance is being shifted to individuals; and the escalating prices of medications and medical technology cry out for change. Seventy-eight million "baby boomers"—those born between 1946 and 1964—will soon be putting enormous stress on the health care system and on Medicare as well. Just the impact of Alzheimer's and other diseases that increase with longevity present a frightening prospect. It behooves us to prepare for this approaching "baby boomer tsunami." The Fall 2006 edition of *EAP Digest* reported that the number of adults above age fifty who use illicit drugs is projected to increase from 1.6 million in 2001 to 3.5 million in 2020 (Watkins 2006). If this proves true, it will create a double burden on society: issues of aging and of illicit drugs.

In 10,000 BC the world had 10 million human inhabitants. By the year AD 1, the number had increased to 300 million. In 1900 we had approximately1.7 billion people on Mother Earth. In 2000 we had slightly over 6 billion. Even though the United States is expected to have modest growth through the year 2020, the world as a whole will dramatically increase in numbers. With the expected doubling of world population every fifty-one years (Tishler 2007), the earth should have about 8 billion residents by 2020, with elderly population more than doubling to more than 1 billion. As we are progressively becoming tied to the global community, the sheer numbers alone will strain our health care system.

The year 2020 will present some very real ethical dilemmas as well. When is "clinically dead" death? Where will cloning lead? How much will humans who are "carbon based" be interfacing with intelligence that is "silicon based"? At what age is it too old to bear children? Where is the balance between "first do no harm" and "we must *do* something?" Should medical resources be rationed? Will we continue to allow the price gouging of the so-called "me too" drugs? It seems as though there are now forces driving the medical industry beyond the professional healer. Until some structural corrections are made, people will be urged to have unnecessary lab tests, encouraged to take some pills that have more side effects than benefits, and advised to seek medical procedures for reasons other than health care. As long as the consumer is coerced by the high-pressure advertising of the pharmaceutical companies and influenced by the media, the system will continue to medicalize normal events and treat common maladies with costly procedures.

In order to maintain high levels of wellness, one must avoid physical decline, not just repair damage. If the doctor returns to being the "teacher," and promotes preventative lifestyle changes, collective wellness will improve. As science develops predictive procedures and genetic profiles, a doctor will be able to use technology as a preventative rather than a curative tool. In 1920 the medical profession knew very little about causes and prevention of disease. The technology we know today did not yet exist. We can only imagine what understandings may be provided by the technology of 2020.

We are going to see increasing application of gene-replacement therapy. Procedures are being developed today to treat serious illness; but gene replacement will eventually be used to boost enzyme levels and hormone production, and even to slow the aging process. Responsible stem cell research will probably proceed with strong guidelines. In 2020 medicine will more readily incorporate the holistic model of wellness. Engaging mind and spirit will progressively become part of medical care. The popular book, *The Power to Heal*, correctly states: "Pressed both by patients and its own advancing technology, medicine will change its focus from treatment to enhancement, from repair to improvement, from diminished sickness to increase performance" (Smolan, Moffitt, Naythons 1990, 192).

As we approach 2020, poverty will likely be a global concern. It is the issue of poverty that keeps babies from being vaccinated, clean water and sanitation from being available, and medical treatment from being provided. Nations are increasingly becoming globally interdependent. In the future, threats to food supply, infectious diseases, natural disasters, terrorism, and other warfare will likely continue. Many of these situations will have significant impact on economic growth and development which further underscores the need for closer links between health systems and other institutions and sectors of our nation and indeed the world.

The bottom line is that the health care is in drastic need of improvement. To use medical nomenclature, the system is not terminal, but its condition is critical. Not only is our health care not measuring up to that of other countries, our costs are swelling even as the number of our uninsured rises.

In 1992 the NIH established the Office of Alternative Medicine. In 1999, the NIH reshaped that agency into the Center for Complementary and Alterative Medicine. With this kind of support and vision from the NIH, integrative and holistic approaches to health are here to stay.

Yet even with systemic strategies and improvement, professional commitment to integrative forms of health care is needed. This author recently heard a psychiatrist say, "There is no more meaning in *depression* than in *diabetes*." Unfortunately, this is a sentiment all too prevalent in the chemical mindset of medicine in the last few years. It is entirely possible that both terms are meaningful. Concerning depression one may ask, "What came first, the chicken or the egg?" That is to ask, has the patient's mood affected the blood chemistry or has imbalanced chemistry affected the mood? Integrative medicine considers the "both/and" as well as the "either/or" scenario. In fact, antidepressants along with psychotherapy, work best. We must, and indeed will, move toward a more integrative form of care in the year 2020.

The Soul of Health—A Summary

To look at this more closely, the Greek word *psyche* refers to the soul. The word *therapist* comes from the Greek word *therapon* which means "one who helps or heals." The Latin equivalent of *therapon* is *ministerium*, from which we get the word *minister*. So, fundamentally, a psychotherapist is a *soul minister*. Now we are being integrative. To proceed with this awareness is to be holistic.

George Lucas in his classic series *Star Wars* gave us a profound spiritual awareness with the phrase, "May the force be with you." There is a spiritual force to health that often goes unrecognized. If only linear intervention is used to address disease, one is not likely to achieve optimal health. Most people are aware of how their emotions may result in a surge of physical responses. Likewise, many who claim a spiritual awareness may still fail to equate this with health.

Wholeness results when we are nurturing all aspects of our being. It is when we embrace the complexities of life—with both its mountaintop and valley experiences—that we begin to participate in the process of becoming whole. Belief, among other things, is a spiritual involvement. One's belief adds power. Certainly we see this power of belief in such things as placebos, hope, expectations and even hypnosis. Perhaps we are only scratching the surface.

Tenzin Gyatso, the 14th Dalai Lama, is one of the most influential spiritual leaders of our time. He is the recipient of the Wallenberg Award, the Albert Schweitzer Award, and the Nobel Peace Prize. His approach to happiness and well-being resembles a mind-science similar to proven techniques of psychotherapy. But, while psychotherapy seeks to *remove* negative mental states and/or bad habits, the Dalai Lama creates well-being not by taking away but by "adding to." His belief is that positive states of mind, like love, compassion, patience, generosity, and kindness, can act as direct antidotes to negative states of mind. Herein, the goal is to achieve healthier coping strategies and amelioration of symptoms. The Dalai Lama says: ". . . if you think seriously about the true meaning of spiritual practices, it has to do with the development and training of your mental state, attitudes, and psychological and emotional state of well-being. . . . True spirituality is a mental attitude that you can practice at any time" (1998, 299).

To engage in life in this fashion is to experience the spiritual dimension of existence—even with life's challenges. When we broaden our paradigm of health to understand that balance is acquired sometimes with multiple forces, we are then on the path of wellness.

There are numerous research studies on the correlation of prayer intervention with improved health. Physician Larry Dossey has written many books documenting this research. His work reveals that prayer can positively affect such things as high blood pressure, wounds, heart attacks, and emotional mood swings. In fact, nothing seemed capable of blocking the healing power of prayer. Indeed, Dossey reached a point where he resolved that failure to use prayer in his practice was equivalent to withholding a necessary potent drug or surgical procedure. In his book *Prayer Is Good Medicine*, Dossey says, "As a physician, I have employed medications and surgical procedures because I know they work. But prayer works, too" (1996, 5).

Homeostasis results when one recognizes that physical is also emotional and that which is emotional is also spiritual. As water becomes solid, then liquid, then gas, our lives pulsate awareness, feelings, and purpose. To ignore our spirit (individually or collectively), is to attain a state of wellness far less than optimal. *May the force be with you* in your pursuit for holistic health and wellness.

References

Badzey, Thomas, ed. 2006. "U.S. Government: Largest Study of Joint-Health Nutrients Impressive." *Health Breakthroughs*. Issue 102.

Barry, John M. 2004. *The Great Influenza*. New York, NY: Penguin Group.

Barry, Patricia 2008. "Discovery: Fatal Drug Errors Soar." *AARP Bulletin*. Vol. 49, No. 7.

Bates, D. 1995. "Incidence of Adverse Drug Events and Potential Adverse Drug Events." *JAMA*, 274.

Centers for Disease Control and Prevention 2004a. "Overweight and Obesity: Obesity Trends: U.S. Obesity Trends 1985-2004." www.cdc.gov/nccdphp/dnpa/obesity/trend/maps/index.htm. (accessed March 23, 2006).

Centers for Disease Control and Prevention. 2004b. "The Health Consequences of Smoking: A Report of the Surgeon General." www.cdc.gov/tobacco/sgr/sgr_2004/index.htm. (accessed March 30, 2006).

Centers for Disease Control and Prevention. 2006a. "Cases of HIV Infection and AIDS in the United States, 2004." www.cdc.gov/hiv/topics/survellance/resources/reports/2004 report/commentary.htm.

Centers for Disease Control and Prevention. 2006b. "What You Should Know about Avian Flu." February 7, 2006. www.cdc.gov/flu/avian/gen-info/facts.htm. (accessed March 23, 2006).

Cohen, Jay S. 2001. *Over Dose: The Case Against the Drug Companies*. New York: Penguin Putnam.

Deyo, Richard A. and Donald L. Patrick. 2005. *Hope or Hype*. New York: Amacom Books.

Dossey, Larry. 1996. *Prayer Is Good Medicine*. New York: HarperCollins.

Ehrbeck, Tilman and Kimberly O'Neill Packard. 2005. "Will Consumer-Driven Health Care Take Off?" *Expert Voices*, Issue 8. Washington, DC: NIHCM Foundation. www.nihcm.org. (accessed March 14, 2006).

Eisenberg, John M. 2001. "Evidence-Based Medicine." *Expert Voices*. Issue 1. Washington, DC: NIHCM Foundation.www.nihcm.org. (accessed March 14, 2006).

Glasser, William. 2003. *Warning: Psychiatry Can Be Hazardous to Your Mental Health*. New York: Harper Collins.

Gyatso, Tenzin (the 14th Dalai Lama) and Howard C. Cutler. (1998). *The Art Of Happiness: A Handbook for Living*. New York: Riverhead Books.

Harris, Margo M. 2005. *Health and Wellness: Illness among Americans*. Farmington Hills, MI: Thomson Gale.

Hoffman, Earl Dirk, Jr., Barbara S. Klees, Catherine A. Curtis. 2005. *Brief Summaries of Medicare & Medicaid*. "Projected Expenditures." Centers for Medicare & Medicaid Services, U.S. Department of Health and Human Services. www.cms.hhs.gov/MedicareProgram Rates/downloads/MedicareMedicaidSummaries2005.pdf. (accessed April 27, 2006).

Jelinek, Pauline. 2006. "Uncle Sam Wants You, but Not if You're a Ritalin-taking, Overweight Generation Y Couch Potato." *Sidney Daily News,* March 13, 2006. Sidney, Ohio: Brown Publishing.

Johnson, Linda A. 2007. "Merck agrees to $4.85 billion Vioxx settlement." *Sidney Daily News* November 10, 2007. Sidney, Ohio: Brown Publishing.

Journal of Employee Assistance. 2006. "Spain Abolishes Smoking at Work as of 2006." Vol. 36, No. 1. Employee Assistance Professionals Association.

Lazarous, J., B. H. Pomeranz, and P. N. Corey. 1998. "Incidence of adverse drug reactions in Hospitalized Patients: A Meta-analysis of Prospective Studies." *JAMA*, April 15, 1998.

Menz, Robert. 1987. "Aiding Those with AIDS: A Mission for the Church." *Journal of Psychology and Christianity*. Batavia, IL: Christian Association for Psychological Studies.

Menz, Robert. 2003. *A Pastoral Counselor's Model for Wellness in the Workplace: Psychergonomics*. Binghampton, NY: Haworth Pastoral Press.

Moore, T. J. 1995. Deadly Medicine: *Why Tens of Thousands of Heart Patients Died in America's Worst Drug Disaster*. New York: Simon and Schuster.

National Institute for Health Care Management. 2000. "Prescription Drugs and Mass Media Advertising, 2000." Washington, DC: NIHCM Foundation. www.nihcm.org. (accessed March 14, 2006).

National Institute for Health Care Management. 2005. "Health Plans Emerging as Pragmatic Partners in Fight Against Obesity." Washington, DC: NIHCM Foundation. www.nihcm.org. (accessed March 14, 2006).

Physician's Desk Reference. 2008. 62nd edition. Stamford, CT: Thomson Healthcare.

Porter, Ray. 1997. *The Greatest Benefit to Mankind: A Medical History of Humanity*. New York: W.W. Norton.

Ramin, Cathyrn Jakobson. 2006. "Valley of the Dulls." *O, The Oprah Magazine*, March, 2006.

Reich, Robert B. 2001. *The Future of Success*. New York: Alfred A. Knopf.

Smolan, Rick, Phillip Moffitt, Mathew Naythons. (1990). *The Power to Heal*. New York: Prentice Hall Press.

Stepnick, Larry. 2003. "Accelerating the Adoption of Preventative Health Services." Washington, DC: NIHCM Foundation. www.nihcm.org. (accessed March 14, 2006).

Strand, Ray D. 2002. *What Your Doctor Doesn't Know about Nutritional Medicine*. Nashville: Thomas Nelson.

Tischler, Henry L. 2007. *Introduction to Sociology*, 9th ed. Belmont, CA: Wadsworth/Thomson Learning.

Watkins, George, ed. 2006. "Aging Baby-Boom Generation May Result in Increased Number of Older Adults Using Drugs." *EAP Digest*, Fall 2006. Employee Assistance Professionals Association.

Weil, Andrew. 1998. *Health and Healing*. New York: Houghton Mifflin.

Weil, Andrew. 2004. *The Natural Mind*. New York: Houghton Mifflin.

Wexler, Barbara. 2003. *Health and Wellness: Illness Among Americans*. Farmington Hills, MI: Thomson Gale.

World Health Organization. 2002. "Health and Sustainable Development: Key Health Trends." Y. von Schirnding and C. Mulholland. www.who.int/media-centre/events/HSD_Plaq_02.2_Gb_def1.pdf. (accessed March 30, 2006).

Chapter 7

Race and Ethnicity

Brian C. Thomas, JD

> And [the people] said, Go to, let us build us a city, and a tower, whose top may reach unto heaven; and let us make us a name, lest we be scattered abroad upon the face of the whole earth.
>
> And the Lord came down to see the city and the tower, which the children of men builded.
>
> And the Lord said, Behold, the people is one, and they have all one language; and this they begin to do: and now nothing will be restrained from them, which they have imagined to do.
>
> Go to, let us go down, and there confound their language, that they may not understand one another's speech.
>
> So the Lord scattered them abroad from thence upon the face of all the earth. . . .
>
> (Genesis 11:4-8, the *Holy Bible*, King James Version)

Scattering mankind abroad certainly prevented Man from climbing toward Heaven. But it may have also marked the beginning of man's long history of racial and ethnic intolerance. One can scarcely read the newspaper or watch the evening news without confronting questions about race and ethnicity. The subject of race permeates the way we think, act, and often with whom we choose to associate. The appearance of the terms "racial profiling" and "ethnic cleansing" in the American lexicon suggests one immutable truth: race and ethnicity matter.

This is understandable. Various studies have shown that race is among the first things that someone notices about another person (DiTomaso 2003). Racial and ethnic characteristics are often easily discernible. Darker or lighter skin tones, eye color, and hair texture are evident even with a cursory glance. Simply identifying physical differences among various races and ethnicities is not, in itself, problematic. However, people may attribute certain characteristics to certain racial or ethnic groups (stereotyping) or treat members of other races and ethnicities differently. It is then that problems often arise.

Racism and ethnic intolerance can take many forms. It can be hidden and barely discernible or it can result in the subjugation of an entire group of people. Perhaps the most troubling aspect of this racial and ethnic intolerance is that intolerant attitudes seem to repeat at regular intervals.

For instance, one can easily draw a parallel between the anti-Japanese sentiments in the early 1940s and the anti-Arab sentiments that emerged following the September 11th terrorist attacks. Both are xenophobic reactions to tragic events. But why does society accept and even promote the harassment of a particular racial or ethnic group as a result of an event? The answer, it seems, is rooted in our intolerant history.

Historical Perspective

The world has long endured a regrettable history of slavery. Slavery has existed, in at least some form, in almost all cultures. Slavery can serve not only to physically bind an entire race of people, its more sinister aim is that it can effectively frustrate that entire group's sense of hope and self-worth. The transatlantic slave trade from Africa to the Americas, which began in the 1500s, epitomized the ultimate racial contradiction. On one hand, America declared its independence in 1776 by proclaiming ". . . that all men are created equal," while at the same time subjecting an entire race of people to unequal and inhuman treatment.

Even after the Emancipation Proclamation abolished slavery in the United States in 1863, racial discrimination continued to pervade the social landscape. The institution of slavery, by its very nature, promoted racist ideologies in the United States; then the discriminatory "Jim Crow" laws, which arose after slavery was abolished, mandated state-sponsored racism and xenophobia for years to come.

Jim Crow's Legacy

Throughout history, governments have sanctioned racism and xenophobia. After the Emancipation Proclamation outlawed slavery, states and municipalities instituted so-called "Jim Crow laws," which codified institutional racism. These laws stratified the populace and reinforced unequal treatment.

For example, the state of Alabama required that all bus passenger stations must have separate waiting rooms or space and separate ticket windows for the white and "colored" races (Alabama 1940). But Jim Crow laws governed more than an individual's ability to travel freely. Some Jim Crow laws governed where and with whom a person could live. Florida, for example, enacted a cohabitation statute in 1903, which provided that "[a]ny negro man and white woman, or any white man and negro woman, who are not married to each other, who shall habitually live in and occupy in the nighttime the same room shall each be punished by imprisonment not exceeding twelve months, or by fine not exceeding five hundred dollars." (Florida 1903, 1). Jim Crow laws even extended into the classroom, where Florida deemed it a criminal offense for teachers of one race to instruct pupils of another race in public schools (Florida State Constitution 1885). Although the Civil Rights Act of 1875 repealed many of these laws, many existed in some form well into the 1960s and 1970s and required judicial intervention to end the legally sanctioned discrimination (Civil Rights Act 1875).

Ending Jim Crow Segregation

Dr. Martin Luther King wrote from a Birmingham jail in 1963, "Any law that uplifts human personality is just. Any law that degrades human personality is unjust. All segregation statutes are unjust because segregation distorts the soul and damages the personality. It gives the segregator a false sense of superiority and the segregated a false sense of inferiority." (King 1963, 5). Under the weight of Dr. King's reasoning, America simply could not support Jim Crow's vast contradiction. That is, one could not celebrate the Fourteenth Amendment's principle of equal protection under the law and, at the same time, treat certain racial and ethnic groups unequally. In light of this contradiction, a number of state and federal courts invalidated these Jim Crow laws, which had not previously been repealed by the legislatures. Following the United States Supreme Court decision in *Brown v. Board of Education*, schools were desegregated and the Civil Rights Movement further dismantled the vestiges of Jim Crow's legacy (Brown 1954). But that invites the obvious question, "How far have we come?"

Current State of Race and Ethnicity

Much progress has been made since the time of Jim Crow, but further progress is not guaranteed. Although segregation is no longer legally sanctioned, some forms of segregation may be resurfacing. According to the Civil Rights Project at Harvard University, the actual desegregation of U.S. public schools peaked in 1988 and, since that time, our schools have become more and more segregated (Kozol 2005). Harvard Law School scholar Charles Ogletree reported that more than eighty-five percent of the schools with overwhelmingly black or Hispanic students are in areas of concentrated poverty (Frankenberg 2003). Many of these minority students will not have access to the same educational opportunities as their white counterparts. In fact, as of 2005, the proportion of black students at majority white schools was at "a level lower than in any year since 1968." (Frankenburg 2003, 6). There may be a number of reasons why schools are increasingly becoming segregated again. Certainly, one explanation for the decline in black students at majority white schools can be explained by the recent state rollback of affirmative action programs.

Dismantling Affirmative Action

On March 6, 1961, President John F. Kennedy issued Executive Order 10925, which created the Committee On Equal Employment Opportunity and mandated that projects financed with federal funds "take affirmative action" to ensure that hiring and employment practices are free of racial bias (Kennedy 1961). Then, on September 24, 1965, President Lyndon Johnson issued Executive Order 11246, which enforced affirmative action for the first time. (Johnson 1965). This executive order required government contractors to "take affirmative action" toward prospective minority employees in all aspects of hiring and employment. Contractors were required to take specific measures to ensure equality in hiring and had to document these efforts. On October 13, 1967, the order was amended to include discrimination on the basis of gender. (Johnson 1967).

The term "affirmative action" is, and will long continue to be, an explosive term. Some equate affirmative action with unfair quotas and systematic discrimination while others believe that affirmative action is a valid means of ensuring opportunities for qualified minority students and job candidates. Because of this dissonance, a prospective student petitioned the United States Supreme Court in 1978. In the landmark case of *Regents of the University of California v. Bakke*, Allan Bakke, a white applicant to the University of California medical school, was twice rejected admission even though some minority applicants were admitted to the school with significantly lower scores (*Regents* 1978). The medical school employed strict affirmative action quotas at the time. Bakke argued that judging him on the basis of his race was a violation of the Equal Protection Clause of the Fourteenth Amendment. In a five-to-four decision, the U.S. Supreme Court ruled that while race was a legitimate factor in school admissions, the use of the kind of inflexible quotas that the medical school used was not.

Although the *Bakke* decision prohibited the use of quotas in college admissions, the affirmative action debate was just beginning. Because of the *Bakke* decision, the term "affirmative action" no longer referred to quotas. Rather, affirmative action programs referred to the attempts to recruit and advance minorities, women, and, in some cases, persons with disabilities. In many instances, affirmative action programs included training programs and other outreach efforts. However, despite the constitutionality of these non-quota programs, a number of states began rolling back their affirmative action programs immediately following the *Bakke* decision.

For example, in 1997, California enacted *Proposition 209*, which stated "The state shall not discriminate against, or grant preferential treatment to, any individual or group on the basis of race, sex, color, ethnicity, or national origin in the operation of public employment, public education, or public contracting." (Proposition 209 1997, 1). Enrollment of black students in California has declined ever since. For example, during the 1970s the University of California's Boalt Hall law school had an aggressive affirmative action plan (Cross 2006). Under this plan, the enrollment of black first year law students reached a high of thirty-nine students in the mid-1970s (Cross 2006). In 1994, the law school had thirty-one such students (2006). In 1996, one year before the affirmative action ban went into effect, twenty were enrolled in the program. However, in 1997, when the affirmative action ban went into effect, the number of black law students admitted to the law school dropped from twenty in 1996 to only one (2006). Overall, the black students admitted to all of the University of California's graduate programs dropped from seventy-seven to eighteen, a seventy-seven percent decrease in one year (2006). Although the enrollment for black students in all graduate programs rebounded to a high of sixteen in 2003, only nine black students enrolled in the law school program in 2006, which is well below the pre-Proposition 209 levels. (2006).

This trend continued at other California graduate schools. In fact, the *Journal of Blacks in Higher Education* released these sobering results:

> In 1994, the year before the affirmative action admissions ban was announced, black enrollments at UCLA law school reached an all-time high of 46. The next year, after the ban was announced, black enrollments dropped by more than one half to 20. In 1997, when the ban took effect, there were only 10 black first-year students at UCLA. By 1999, black enrollments reached a level not seen since the early 1960s. In 1999, only three black first-year law students enrolled in UCLA. (Cross 2006, 1)

The rollback of affirmative action programs did not occur only in California. The Florida legislature approved the "One Florida Initiative," which was designed to end affirmative action (Saunders 2000).

In addition to California and Florida, other states sought to ban affirmative action programs. In a highly publicized case, several students challenged the University of Michigan affirmative action programs in 2000. Two separate lawsuits were filed, one by students applying to the University of Michigan undergraduate school and one by students applying to the University of Michigan law school (Gratz 2003). The university argued that the affirmative action programs serve a compelling interest by providing educational benefits derived from a diverse student body in the same way that certain preferences are granted to children of alumni, scholarship athletes or other groups for reasons deemed beneficial to the university. A federal judge ruled that the use of race as a factor in admissions at the University of Michigan undergraduate school was, in fact, constitutional (Gratz 2003).

Then, in 2003, the U.S. Supreme Court heard the Michigan law school and undergraduate cases. In a six-to-three decision, the Court held that the formula used by the University of Michigan undergraduate admissions program was unconstitutional and had to be modified. The undergraduate program used a point system to rate students and awarded additional points to minority students. The Supreme Court invalidated the program as it was constructed because it did not provide adequate individualized consideration.

However, in a five-to-four decision, the Supreme Court upheld the University of Michigan law school's affirmative action program. The Court ruled that, under federal law, state-operated law, business, medical, and other professional schools could consider race as a positive factor in the admissions process (Grutter 2003). The Court further held that race can be one of the many factors considered by colleges when selecting their students because it furthers "a compelling interest in obtaining the educational benefits that flow from a diverse student body." Although this ruling appears to validate the use of affirmative action programs, it actually provides only that federal law cannot prohibit schools from using affirmative action as a positive factor in admissions. Individual states are free to ban affirmative action programs if they choose to do so.

Many states decided to do just that. In November 2006, Michigan voters approved *Proposal 2*, which prohibits affirmative action preferences. The proposal passed by a fifty-eight to forty-two percent margin (Slevin 2006). While the outcome of this law is not yet known, it is reasonable to assume that the trend of lower enrollment of black students will continue. The invalidation of affirmative action programs will likely have far-reaching impact.

William J. Edwards wrote, "Education is the source of all we have and the spring of all our future joys." (Ealy 1997, 1). However, these future joys may be out of reach for some racial and ethnic minorities if they are denied access to the spring. Moreover, if racial and ethnic minorities do not receive the same access to education as the majority population, it follows that they will not receive the same access to select jobs and career opportunities. The trend towards invalidating affirmative action programs may continue in other states. One cannot accurately anticipate how the dismantling of affirmative action programs will affect the minority populations. The trend toward prohibiting affirmative action programs offers a glimpse of how our attitudes about race and ethnicity are rapidly changing. However, our collective attitudes about race and ethnicity may not be changing as fast as the population itself.

Increases in Racial and Ethnic Diversity in the United States

Countless studies have shown that America's population is becoming increasingly diverse. In 1980, seventy-four percent U.S. children under the age of eighteen were white, whereas fifteen percent, black, nine percent, Hispanic and two percent, Asian. Contrast that with projections for the year 2020, where fifty-five percent U.S. children under the age of eighteen are estimated to be white, seventeen percent, black, twenty-two percent, Hispanic, and six percent, Asian (Campbell 1996). These projections confirm the trend of the United States' becoming a more diverse nation. Although these population projections point to a rapid demographic shift, our attitudes toward race and ethnicity do not appear to be changing quite as rapidly.

For example, the Opinion Research Corporation conducted a telephone poll from December 5-7, 2006 for CNN. 1,207 adult Americans responded to the poll (CNN 2006). Of the participants, 328 were black and 703 were white non-Hispanics. During the telephone poll, Opinion Research asked the respondents a number of questions about racial and ethnic attitudes. Eighty-four percent of black respondents considered racial bias as very serious or somewhat serious, whereas sixty-six percent of white respondents considered racial bias a very or somewhat serious problem. While eighty-seven percent of black respondents indicated that they would approve if their son or daughter married someone of a different race or ethnic group, only sixty-nine percent of white respondents said they would approve if their son or daughter did so. (2006) What do these results tell us?

The fact that most black and white respondents overwhelmingly view racial bias as a serious issue in America but, at the same time, overwhelmingly consider themselves not to be racially biased, is inherently contradictory. This contradiction may explain some of the racial tensions that exists today. Apparently, many view racism and xenophobia as a serious problem, as the study suggests, but not *their* problem. It is a problem that someone else needs to solve. In other words, why would I change my attitudes about race? I am not racially biased. Yet, the poll results suggests otherwise.

While eighty-six percent of Whites surveyed denied racial bias, only sixty-nine percent would approve if their son or daughter married someone of a different race or ethnic group. The difference between those numbers is an example of racial bias. Attitudes about race and ethnicity do not exist in a vacuum. The positive and negative attitudes that individuals have about members of other racial and ethnic groups often shape their interactions with that group.

Are Attitudes Toward Race/Ethnicity Changing?

Attitudes about race and ethnicity have certainly evolved over time. There are few places where the evolution of racial and ethnic attitudes is more evident than in the sporting arena. In many ways, sports serve as a microcosm of society. Jesse Owens challenged the notion of the racial superiority of German *Aryans* during the 1936 Olympics when he won four gold medals and set Olympic records in three of the four events. Jackie Robinson erased the color line in Major League baseball in 1947, several years before today's Civil Rights movement began. Moreover, in 2007, Indianapolis Colts head football coach Tony Dungy changed attitudes about the ability of black NFL coaches, when he won Super Bowl XLI in a game in which both teams featured black head coaches.

However, sports also demonstrates how attitudes about race and ethnicity have remained unchanged. For example, in 1988, professional football analyst and oddsmaker Jimmy "The Greek" Snyder compared the physical attributes of black and white athletes during a television report. In the interview, Snyder said that the black athlete had been bred since the Civil War to be bigger and stronger. Snyder further commented that black athletes have bigger thighs which allow them to run faster and jump higher (Shapiro 1988). The CBS television network later fired Snyder for his remarks. But Snyder's views on race, while shocking, illustrate an important point about our collective attitudes. When one attributes a particular characteristic to a certain race, it minimizes the individuality of each member of that race.

Although everyone does not share Snyder's stereotypical view of the world, there is ample evidence to suggest that many do. For example, ten years after Snyder's comments, Hall of Fame professional football player Reggie White addressed the Wisconsin Legislature and shared his views on race and ethnicity. White remarked that each race has certain gifts and talents. White said that blacks are gifted at worship and celebration. Whites are good at organization and business. Hispanics are gifted in family structure. Asians are inventive and Indians are gifted in spirituality (*CBS Sportsline* 1998).

White's comments, like those Snyder made ten years earlier, reduce members of a particular race to nothing more than a few representative characteristics. By doing so, White mistakenly assumes an impossibility—that all members of the racial group necessarily share immutable characteristics. Although White attempts to attribute one positive characteristic to each of racial groups he named, many attribute negative characteristics to members of particular racial groups. Moreover, if whites are considered organized and good in business and blacks are gifted at worship and celebration, what happens when equally qualified white and black candidates apply for the same job? Is the white candidate going to be given the edge in interview because he is "naturally" more organized than the black candidate?

While some may dismiss White and Snyder's comments as vestiges of the past, recent events suggests that these attitudes continue to exist. On November 20, 2006, black sports announcer and Hall of Fame football player Michael Irvin found himself embroiled in a firestorm when he commented about the white Dallas Cowboys quarterback Tony Romo. Irvin in essence implied that Romo's ancestors practiced selective breeding (Kerasotis 2006).

Irvin, like Snyder, held the stereotypical belief that black athletes are athletically more gifted than white athletes. The danger of this association is easily apparent. Snyder and Irvin, and anyone who shares their stereotypical beliefs, will maintain different expectations for white athletes as they will for black athletes. As Reggie White's statements indicate, these stereotypical beliefs are not isolated to the sporting arena.

White and others like him will always expect members of a particular group to behave in a certain prescribed way. This will inevitably cause tension when members of a racial group do not act in a way that is consistent with the stereotype. These stereotypes are not particularly problematic if those who hold the stereotypes do not act on their beliefs. However, problems can arise when one who subscribes to rigid stereotypes acts upon his or her stereotypical beliefs.

Transforming Biased Attitudes Into Action

Racism can be characterized as the belief that race is the primary determinant of human traits and capacities and that those racial differences produce an inherent superiority of a particular race. Like racism, stereotyping is a mental conception, which occurs when one develops an oversimplified opinion, affective attitude, or uncritical judgment about members of a particular group. However, while stereotyping involves only a mental conception, racism generally involves some form of acting upon one's prejudices and biases.

Of course, racism does not operate in a vacuum. It often invades our most basic daily interpersonal interactions. For example, the City of Cincinnati, Ohio commissioned the Rand Corporation to conduct a study to analyze police data for possible signs of racial profiling. (Korte 2006). Rand researchers analyzed 335 police videos taken by dashboard mounted cameras in police cruisers. The report found, among other things, that while walking away from a traffic stop of a white driver Cincinnati police officers are more likely to say "Have a nice day" or "take care." Officers are more likely to tell black drivers, "You're getting off easy," or "the ticket could have been twice this amount." The manner in which the officers end a traffic stop speaks volumes about how the officer perceives the drivers. Concluding a traffic stop with the phrase "take care" or "have a nice day" connotes a sense of respect for the driver. In contrast, admonishing a driver that the fine could have been more severe or that the driver is getting off easy suggests that the driver should respect the officer.

Although the Rand researchers found no statistical evidence that the police officers engaged in systematic racial bias when comparing black and white drivers in similar areas and circumstances, the report noted that aggressive policing in high-crime areas may lead some to believe that police treat blacks and whites differently. That is, on average, more African Americans live in high crime neighborhoods. The report noted that people in higher crime areas are more likely to be pulled over, face tougher interrogations and be subject to longer searches than drivers in lower crime areas.

While the aggressive policing in some high crime areas could lead to a perception that police treat blacks and whites differently, this rationale does not explain another startling result. Several Cincinnati police officers pulled over significantly more black drivers than their peers working similar assignments. One explanation for the variance is that these officers are acting upon their innate belief that black drivers are more likely than white drivers to break the law. Because they believe that black drivers are more likely to break the law, they are more apt to stop black drivers. This trend, which has often been described as racial profiling, has infiltrated our culture in various ways.

Profiling the Justice System

In a report entitled, "Driving While Black: Racial Profiling on Our Nation's Highways," University of Toledo College of Law Professor David A. Harris described racial profiling this way:

> Racial profiling is based on the premise that most drug offenses are committed by minorities. The premise is factually untrue, but has nonetheless become a self-fulfilling prophecy. Because police look for drugs primarily among African Americans and Latinos, they find a disproportionate number of them with contraband. Therefore, more minorities are arrested, prosecuted, convicted, and jailed, thus reinforcing the perception that drug trafficking is primarily a minority activity. This perception creates the profile that results in more stops of minority drivers. At the same time, white drivers receive far less police attention, many of the drug dealers and possessors among them go unapprehended, and the perception that whites commit fewer drug offenses than minorities is perpetuated. And so the cycle continues. (Harris 1999, 2)

Professor Harris' analysis may explain the disproportionately high number of non-whites who are apprehended for drug offenses. That is, racial profiling may result in increased arrests and incarcerations among non-whites precisely because police target non-whites at a greater rate than whites who are also committing the same crimes.

Unfortunately, the disproportionate treatment of whites and non-whites is not confined to traffic stops. The vast disparity in the incarceration rate among blacks, whites, and Hispanics suggests that the attitudes on race may find their way into all levels of the American judicial system. For instance, in 2000, the population of children aged ten to fifteen in New York was 56.3% white, 18% black, 17.8% Hispanic (W. Haywood Burns Institute 2000). However, the detention rate for that same group was 60.3% black, 20.4% Hispanic and 17.7% white. In 2002, Georgia's population of children aged ten to seventeen was 59% white, 34% black, 5% Hispanic (2002). However, youths confined in detention facilities were 63% black, 33% white, and 4% Hispanic. Finally, in 2003, California's youth population between the ages of ten and seventeen was 45% Hispanic, 35.8% white, and 8.2% black (2003). As with Georgia and New York, California's detention rate for non-whites was significantly higher than for the white population, with an incarceration rate of 46.7% Hispanic, 25% black, and 24.2% white youths (2003).

These results suggests that geography does not change the trend. All across America non-whites are incarcerated at a much higher rate than whites. Although one can argue that non-whites are incarcerated at a higher rate than whites because they commit more crimes than whites, such thinking ignores Professor Harris' observation that non-whites are stopped and incarcerated at a much higher rate than whites because the police target non-whites for traffic stops. Moreover, Professor Harris' findings are consistent with the Rand researchers' findings that certain police officers target significantly more blacks than their peers working similar assignments.

Moreover, it is absurd to assume that people check their beliefs at the courthouse steps. The same beliefs that could lead a police officer to stop more blacks than whites may also motivate a juror to impose a harsher sentence on a black or Hispanic defendant. The statistics support this fact. In 1999, blacks constituted approximately 13% of the country's drug users, but 37% of people arrested on drug charges, 55% of those convicted, and a staggering 74% of all drug offenders sentenced to prison were black (Harris 1999).

Given these statistics, it is not surprising that attitudes concerning the police and the justice system are drawn, in large part, along racial lines. For example, in February 1999, the researchers at the Social Research Laboratory and Department of Criminal Justice at Northern Arizona University conducted a statewide telephone survey of 433 randomly selected Arizonans (Wonders 1999). The research laboratory compared the results to similar studies that it conducted in 1994 and 1997. The study found that while most Arizona residents had a favorable impression of their local police department, "people of color are twice as likely as whites to have an unfavorable impression of their local police (eighteen versus nine percent)" (Wonders 1999, 2). The report revealed the following differences in how whites and non-whites viewed the police:

> Almost two-out-of-three whites (62%) believe that the police are their friends, while only 44 percent of people from other races share this belief. Half the people of color in the study (49%) say the police are neither their friends nor their enemies, while 35 percent of whites share this view.
>
> A majority of whites in Arizona (57%) say the police treat all people, regardless of race, the same. Less than half of people from other races in Arizona believe this to be true (46%). Thirty-five percent of people

> from other races believe the police are tougher on people of color than on whites. Twenty-three percent of whites share this view.
>
> Perceptions of police abuse appear to be widespread, according to study findings. One-third of people of color in the study (33%) and 19 percent of whites say they personally know someone who has been physically mistreated or abused by the police. (Wonders, 1999, pp. 2-3)

These statistics confirm an unfortunate truth. Members of different racial groups may collectively have different views of police and the judicial system. This is problematic for several reasons. The attitudes that members of different racial groups have about the police will affect how they interact with the police and how the police interact with them during routine encounters. These divergent attitudes may affect whether someone in a racial or ethnic minority wants to pursue a career opportunity in law enforcement. These attitudes may also affect whether police departments ultimately hire members of racial and ethnic minorities. This is by no means an exhaustive list of the problems that can be caused by the divergent views that different racial and ethnic groups have about the police and justice system. However, it serves as a reminder that society must tackle a number of issues as it moves forward toward the year 2020 and beyond.

Future Trends—2020

Society has made great strides in furthering racial and ethnic tolerance. However, as demonstrated above, society still has a long way to go before racial and ethnic intolerance is no longer an issue. Hall of Fame baseball catcher Yogi Berra once said, "If you don't know where you are going, you will wind up somewhere else." (Eller 1997, 1) Although artlessly stated, Berra's sentiment rings true today. Society must address important racial and ethnic issues such as affirmative action, racial profiling, ethnic intimidation, disproportionate incarceration rates of minorities, and disproportionate access to health care and jobless rates among racial and ethnic minorities. If society fails to address these and other issues affecting racial and ethnic minorities, those issues will certainly remain in the year 2020.

The issues must be addressed for another reason. The population is becoming increasingly diverse. This increased diversity will force people to confront racial and ethnic issues firsthand. Society can choose to view this increased diversity as an obstacle or as an opportunity. As Dr. Martin Luther King, Jr. once said, "Let us all hope that the dark clouds of racial prejudice will soon pass away and the deep fog of misunderstanding will be lifted from our fear-drenched communities, and in some not too distant tomorrow the radiant stars of love and brotherhood will shine over our great nation with all their scintillating beauty." (King 1963, 15).

Dr. King's image of a racially diverse and tolerant future provides an ideal vision. It offers a glimpse of what life might be like if we can transcend the racial and ethnic animosities that hold us, as a society, from truly realizing our full potential. As discussed in this chapter, society has made great strides towards making this vision a reality. However, society still has work to do before it can dispel the clouds of racial and ethnic prejudices.

It is no surprise that racial and ethnic intolerance, hatred, prejudice, discrimination and inequality offend our collective sensibilities. These attitudes directly contradict the universal philosophy of loving one's neighbor. As we move toward the year 2020, racial and ethnic tolerance and understanding must be viewed as more than an unattainable ideal. It should be viewed as an exceedingly important goal in our increasingly diverse landscape.

Society is at a crossroads. While the dismantling of Jim Crow's legacy signals a move in the right direction, racial profiling, the disparate incarceration rates for racial minorities, and the rollback of educational opportunities signal a retreat towards our intolerant past. One can only hope that society continues to move toward Dr. King's ideal vision because the alternative is simply too frightening to imagine.

References

ALA. CODE Title 48, § 301(31a-c) (1945), as amended.

Brown v. Board of Education, 347 U.S. 483 (1954).

Cambell, Paul. 1996. "Population Projections for States by Age, Sex, Race, and Hispanic Origin: 1995-2025." Population Projections Branch, Population Division, U.S. Bureau of the Census. October 1996. www.census.gov/population/www/projections/pp147.html. (accessed October 9, 2006).

CBS *Sportsline Wire Reports*. "Packers' Reggie White Spouts Insane Extremist Bigotry and Hate." March 25, 1998. www.skeptictank.org/white.htm. (accessed January 11, 2007).

Civil Rights Act of 1875, 18 US Statutes at Large, Vol. XVIII, 335 (Act of Mar. 1, 1875).

CNN Poll. "Most Americans See Lingering Racism—in Others." December 12, 2006. www.cnn.com/2006/US/12/12/racism.poll/index.html. (accessed December 29, 2006).

Cross, Theodore. 2006. "How State Bans on Race-Sensitive Admissions Have Damaged Black Enrollments in Professional Schools." *The Journal of Blacks in Higher Education*. www.jbhe.com/features/51_professional_schools.html. (accessed February 28, 2007).

DiTomaso, Nancy. 2003. "Race: The Power of an Illusion." www.pbs.org/race/000_About/002_04-background-03-07.htm. (accessed February 8, 2007).

Ealy, Lenore. 1997. "William Edwards: Education crusader in the rural South." Profiles in Citizenship. www.hoover.org/publications/policyreview/3573302.html. (accessed February 28, 2007).

Eller, Suzanne T. (2007). "Real Quotes." www.daretobelieve.org/gpage4.html. (accessed February 28, 2007).

Fla. Stat. Ann. § 798.05 (1903), repealed.

Fla. Constitution Article XII, section 12 (1885), repealed.

Frankenberg, Erica, Chungmei Lee, Gary Orfield (2003). "A Multiracial Society with Segregated Schools: Are We Losing the Dream?" The Civil Rights Project Harvard University. January, 2003. www.civilrightsproject.harvard.edu.

Gratz v. Bollinger, 539 U.S. 244 (2003).

Grutter v. Bollinger, 539 U.S. 306 (2003).

Harris, David. 1999. "Driving While Black: Racial Profiling on Our Nation's Highway." American Civil Liberties Union Special Report. www.aclu.org/racialjustice/racialprofiling/15912pub19990607.htm. (accessed January 11, 2007).

Johnson, Lyndon B. 1964. Executive Order No. 11246, C.F.R. 339 (1964-1965), reprinted as amended in 42 U.S.C. § 2000e (2000).

Johnson, Lyndon B. 1967. Executive Order No. 11246, C.F.R. 339 (1964-1965), reprinted as amended in 42 U.S.C. § 2000e (2000).

Kennedy, John F. 1961. Executive Order No. 10925 26 Fed. Reg. 1977 (Mar. 6, 1961).

Kerasotis, Peter. 2006. "Why Didn't Racial Remarks Sink Michael Irvin?" *Florida Today*. November 28, 2006.

King, Jr., Martin Luther. 1994. "Letter from a Birmingham Jail" (First Edition). New York: HarperCollins.

Korte, Gregory. 2007. "Officers' Farewells Vary By Race." *Cincinnati Enquirer*. January 4, 2007. Cincinnati, Ohio: Gannett Publishing.

Kozol, Jonathan. 2005. "Overcoming Apartheid." www.thenation.com/doc/20051219/kozol. (accessed February 19, 2007).

Proposed Amendment to Article I § 31, California Constitution. CAL. CONST. art. I § 31 (adopted 1996).

Regents of the University of California v. Bakke .1978. 438 US 265.

Saunders, Jim. 2000. "Regents unanimous for One Florida." *The Florida Times-Union*. www.jacksonville.com/tu-online/stories/021800/met_2171246.html. (accessed February 28, 2007).

Slevin, Peter. 2006. "Court Battle Likely on Affirmative Action: Michigan Voters Approved Ban, but Opponents of the Measure Persist." *The Washington Post*. November 18, 2006.

Shapiro, L. 1998. "'Jimmy the Greek' says Blacks are Bred for Sports," *The Washington Post*. January 16.

W. Haywood Burns Institute for Juvenile Justice Fairness and Equity. 2003. California Relative Rate Index data. www.burnsinstitute.org/dmc/ca/.

W. Haywood Burns Institute for Juvenile Justice Fairness and Equity. 2002. Georgia Relative Rate Index data. www.burnsinstitute.org/dmc/ga/.

W. Haywood Burns Institute for Juvenile Justice Fairness and Equity. 2000. New York State Relative Rate Index data. www.burnsinstitute.org/dmc/ny/.

Wonders, Nancy. 2001. "Arizona Crime Rates Steady Over Past Two Years; Support for Education and Prevention Programs to Fight Crime Remains Strong; More Police Wanted in Wake of Terrorist Attacks." Northern Arizona University Social Research Lab. www4.nau.edu/srl/News/10-08-01.pdf. (accessed February 28, 2007).

Chapter 8

Technology

GWENDOLYN OGLE, PH.D.

> Whatever our attitude and understanding it is undeniable that technology is now a major determinant in most human activities, expectations and even beliefs. It has virtually unlimited potential for good; wrongly used it could, as we are so often reminded, bring disaster to us all.
>
> (Cardwell 1995, 3)

Technology Defined

Ask the average person what they think of when they hear the word *technology*, and most will reply, "Computers." Machinery like the computer has been a major component of the technology thrust that has occurred over the last several decades. A technology is actually defined as the practical application of knowledge especially in a particular area, according to the Merriam-Webster Online Dictionary. Therefore, technology can be a tool or a process. What makes technology uniquely human is that tools evolved (and continue to evolve) using the application of recorded knowledge, from oral traditions to the most complex communications systems of today (Kurzweil 1999). This technological evolution is happening at an exponentially accelerating pace.

The word itself was coined in the seventeenth century (Cardwell 1995). Technology comes from the Greek *tekhne* ("craft" or "art") and *logia* ("the study of") (Kurzweil 1999). Since Francis Bacon in the 1600s, it has been commonly accepted that the purpose of science and technology is to improve the human condition (Wishard 2001). Technologies have been developed to automate, streamline, conquer, out-maneuver, simplify, and otherwise "improve upon" the current status quo. There is a life cycle to technology, with seven distinct stages: the precursor or "dream" stage, invention, development, maturity, pretenders (competitors with "improvements"), obsolescence, and antiquity (Kurzweil 1999).

Technology is inarguably tied to science, but there are distinctive differences between the two. Cardwell (1995) emphasizes that *purpose* marks the main distinction between the two. Scientists begin with a theory or objective, which might be vastly different from the resulting outcome. Technologists (and inventors), by contrast, almost always have in mind a concrete outcome. "At the heart of technology lies the ability to recognize a human need, or desire (actual or potential) and then to devise a means—an invention or new design—to satisfy it economically" (Cardwell, 490). That being said, technology innovations are often evolutionary, in that they are continually improved upon and utilized for new purposes.

Cardwell notes that science and technology are so tied, that some boundaries are obscured "to the point of non-existence" (487). "[T]he role of purpose notwithstanding, in the last resort technology and science are aspects of the same thing. They constitute the inseparable procedures by which we attempt to understand and to control the natural world for the benefit and, ultimately, for the survival of humanity" (513).

As technology permeates society and culture, what, if any, are the moral responsibilities of its end-users? Further, what are the moral responsibilities of scientists and technologists engaged in the design and development of technologies? (Cardwell 1995). These are the questions that we will attempt to address as we discuss the social and spiritual visions of technology in the year 2020.

Historical Perspective

Technologies were utilized long before written history would document their implementation. For example, technology was in use when prehistoric cavemen chipped stone arrows for hunting. What distinguishes humans from other intelligent primates is that humans are tool-makers, not merely tool-users (Cardwell 1995).

> *Homo sapiens* are not very different from other advanced primates in terms of their genetic heritage. Their DNA is 98.6 percent the same as

> the lowland gorilla, and 97.8 percent the same as the orangutan. *Homo sapiens* are distinguished by their creation of technology. Technology goes beyond the mere fashioning and use of tools. It involves a record of tool making and a progression in the sophistication of tools. It requires invention and is itself a continuation of evolution by other means . . . the technologically more advanced group ends up becoming dominant. This trend may not bode well as intelligent machines themselves surpass us in intelligence and technological sophistication in the twenty-first century." (Kurzweil 1999, 14-15)

Perhaps the two most important of all ancient technological innovations were metal working and the creation of meaningful symbols—numeric and alphabetic (Cardwell 1995). No one can argue that these are critical foundations upon which much of future science and technology would be built. There came a point in history when the technologies being developed were no longer just for aiding in physical tasks, but for mental tasks as well. Babbage recognized this turning point, what some might recognize as the precursor to modern artificial intelligence.

The list of technologies has certainly grown exponentially in the last few centuries. Just over 100 years ago was the very first flight; now air travel is taken for granted. Computers were once large enough to fill a room, but in just decades they have become as small as a book, yet faster and with greater capacity.

Leadership in the technology movement has changed hands over the last few centuries. Southern Germany and northern Italy led prior to 1600, western Europe and the United States, between 1700 and 1900. Japan and southeast Asia have taken the lead, particularly since 1945 (Cardwell 1995). Cardwell writes that a simple glance around one's home or garage will confirm this.

Five paradigms have existed regarding the exponential growth of computing, the current (fifth) paradigm being Moore's Law on Integrated Circuits. Moore's Law states that every two years, you can get twice the number of transistors into an integrated circuit, thus doubling its speed. Moore's Law was created by Gordon Moore (inventor of the integrated circuit) and has been remarkably accurate since it was stated over forty years ago. This trend is expected to last for approximately fifteen more years, until transistors are just a few atoms thick. Then a new theorem will replace Moore's Law, as is inherent in the Law of Time and Chaos (Kurzweil 1999).

The Law of Time and Chaos states, "In a process, the time interval between salient events (that is, events that change the nature of the process, or significantly affect the future of the process) expands or contracts along with the amount of chaos." The inverse sublaw, the Law of Accelerating Returns, states "As order exponentially increases, time exponentially speeds up (that is, the time interval between salient events grows shorter as time passes.)" The Law of Accelerating Returns (applied specifically to evolutionary processes) explains the exponential development of technology (an evolutionary process) quite well. As we create order, and thus minimize chaos, the evolution of technology speeds up (Kurzweil 1999, 29-30.) In other words, technology builds on its own increasing order, at an increasing speed.

Current State of Technology

William Van Dusen Wishard suggested in his 2001 address, "Between Two Ages" (given to the Coudert Institute), that we are probably living at "the most critical turning point of human history," and that the "next three decades may be the most decisive thirty-year period in human history" (2001, 231). He says, "We're in what the ancient Greeks called Kairos—the 'right moment' for a fundamental change in principles and symbols" (226). Wishard also says,

> what we're experiencing is not simply the acceleration of the pace of change, but the acceleration of acceleration itself. In other words, change growing at an exponential rate. The experts tell us that the rate of change doubles every decade; that at today's rate of change, we'll experience 100 calendar years of change in the next twenty-five years; and that due to the nature of exponential growth, the twenty-first century as a whole will experience almost one thousand times greater technological change than did the twentieth. (229)

There were 3,891,901 technology patents applied for between January 1, 1963 and December 31, 2005, according to the Patent Technology Monitoring Division report profiling utility patent activity in all technology areas (United States Patent and Trademark Office, 2006). Ian Neild, co-author of "A Timeline for Technology" (2005) says, "Human technology has moved from the first flight to flying to the moon in around 60 years—which was a remarkable achievement . . . [In the next sixty years] we will see nanotechnology and biotechnology making impacts on our life that might seem like magic to us but will be quite normal to our children's children."

Coupled with this increasing pace of technological change is an interesting shift: from specialized sciences to scientific, technological convergence.

> Recently, a small but influential network of scientists, engineers, and scholars has coalesced with the aim of promoting technological convergence. These Convergenists have focused on the synergistic unification of the *NBIC* fields: nanotechnology, biotechnology, information technology and . . . cognitive science. Convergenists advocate aggressive research in cognitive science, including computational neuroscience, to understand how the human brain actually creates the mind, and thus how to emulate it. (Bainbridge 2006, 28)

Many see the future riding on the convergence of these four fields. So what are the potentials from these fields?

Nanotechnology is defined as "the science of developing materials at the atomic and molecular level in order to imbue them with special electrical and chemical properties" (*Computer Desktop Encyclopedia* 2006). Nanotechnology is expected to make a significant contribution to the fields of computer science, biotechnology, manufacturing and energy. "Envisioned are all kinds of amazing products, including extraordinarily tiny computers that are very powerful, building materials that withstand earthquakes, advanced systems for drug delivery and custom-tailored pharmaceuticals as well as the elimination of invasive surgery, because repairs can be made from within the body" (2006).

Nanotechnology employs devices typically less than 100 nanometers in size. A nanometer is one *billionth* of a meter—far smaller than the head of a pin which is 1 *million* nanometers wide (Austin 2004). Nanotechnology builds upon several highly specific disciplines, including physics, chemistry, engineering, and quantum mechanics. Nanotechnology, bionanotechnology and nanomedicine describe the merging fields of nanotechnology and medicine. Scientists have already discovered microbes that act as a motor (spinning a rotor), algae that act as packmules, bacteria that lay electrical conduits, and cells that pump (Baker 2006).

A national survey conducted by North Carolina State University (Austin 2004) found that most Americans hold a positive view about nanotechnology and its potential benefits, despite lacking concrete understanding and knowledge of the topic. (Eighty percent indicated they had heard from "little" to "nothing" about nanotechnology.) Dr. Michael Cobb, assistant professor at North Carolina State who analyzed the results of the survey he designed, feels this reflects American's positive views of science in general. Survey respondents were asked to choose the most important benefit (from a list of five options), and fifty-seven percent selected better detection and treatment methods for diseases. This was followed by cleaner environment (sixteen percent), national security and defense capabilities (twelve percent), improved human physical and mental abilities (eleven percent), and cheaper, longer-lasting consumer products (four percent). Potential risks cited were privacy issues, nanotechnology-inspired arms race, breathing nano-sized particles (toxicity), job loss and economic disruption, and self-replication of nanorobots (Austin 2004).

Biotechnology is "the use of biological processes, as through the exploitation and manipulation of living organisms or biological systems, in the development or manufacture of a product or in the technological solution to a problem" (*The Columbia Electronic Encyclopedia* 2003). There is evidence of use of biotechnology 10,000 years ago when humans first implemented selective breeding.

What are the potential benefits of biotechnology? According to the Convention on Biological Diversity website (2005),

> Genetic engineering promises remarkable advances in medicine, agriculture, and other fields. These may include new medical treatments and vaccines, new industrial products, and improved fibers and fuels. Proponents of the technology argue that biotechnology has the potential to lead to increases in food security, decreased pressure on land use, sustainable yield increase in marginal lands or inhospitable environments and reduced use of water and agrochemicals in agriculture.

There is a current societal and religious debate regarding stem cell research. Those who believe life begins at the fertilization of the egg of course see a problem with stem cell research. The embryo's development is stopped at a point at which stem cells can be harvested, and to some this is a death, or worse, a murder. Spiritual and religious beliefs certainly play a role in the politics of implementing new, unprecedented biomedical research.

Artificial intelligence (AI) is defined as "the ability of a machine to perform tasks thought to require human intelligence" (*Britannica Con-*

cise Encyclopedia 2006). The term was coined by John McCarthy in 1956, when he organized a workshop that brought together people from diverse fields to form a consensus on the nature and direction of the new field of study that he had named (Hogan 1997). AI takes cognitive science to a whole new level, for one could not possibly create a machine that simulates human thinking, without a better understanding of how humans think.

> Although pseudo-intelligent machinery dates back to antiquity, the first glimmerings of true intelligence awaited the development of *digital computers* in the 1940s. AI, or at least the semblance of intelligence, has developed in parallel with computer processing power, which appears to be the main limiting factor. (*Britannica Concise Encyclopedia* 2006)

Early applications of "artificial intelligence" such as playing chess and solving mathematical problems are now seen as trivial, even antiquated, compared to visual *pattern recognition*, complex decision making, and robotics seen in today's AI.

Nonetheless, AI is still inferior to human intelligence in that computers can do only what they are programmed to do. Machines have the irritating tendency "to do what they're told instead of what they are supposed to know was meant" (Hogan 1997, 3). They lack the methods that humans use to scale down an otherwise enormous task. In other words, a computer will search the whole of its memory to come to an answer, whereas a human will know to eliminate extraneous information and focus its attention on that which will most likely give an answer. As Hogan puts it, "to make machines appear reasonably smart in the ways we humans take pride in being smart, one must use logically defective shortcuts," known as 'heuristics' (1997, 288).

Like newly-made machines, humans come into this world with no knowledge (other than basic instinctual knowledge.) The *experiences* that a human has throughout life, combined with the knowledge programmed in by parents, books, teachers, and so on, and the ability to learn from these experiences, are what constitute our intelligence (Hogan 1997). It is in programming these *processes* that those in the field of AI have difficulty, and what AI research is all about.

Paradoxically, despite AI's early success (when machines so "easily" handled tasks that seemed cumbersome for the human mind such as advanced calculations and code-cracking), AI would later encounter great difficulty in performing tasks we consider "mindless." The processes that seem instinctual to us, inherent and simple, have proven quite difficult to teach a computer. The problem may be an inability to see the forest for the trees. How do we program that which is so inherent that the procedural steps elude us? Hogan terms this dilemma "What teachers don't have to teach" (1997, 231).

When discussing artificial intelligence, one cannot overlook the fact that to truly replicate human intelligence, machines must be programmed to share our capacity for self-awareness and to feel our spectrum of emotions (Hogan 1997). Although Hogan theoretically concludes that this could happen—why shouldn't a different form of system, created a different way, be capable of doing what more than five billion other electromechanical systems operating within the laws of physics are doing?—he doesn't predict that it is likely. Hogan predicts that whatever form of self-awareness develops as machines become infinitely more complex, it will be quite unlike our own. Over time, this could prove cataclysmic, according to Vernor Vinge (1993). He describes "Singularity," a point at which our role as the dominant species is over, relinquished to the machine.

Commenting on whether or not humans will be the most intelligent beings by the end of the twenty-first century, Kurzweil concludes that the answer depends on how we define *human*. He writes, "The primary political and philosophical issue of the next century will be the definition of who we are" (1999, 2). Hirsch writes, "Once basic technical issues are understood, policy issues become ethical and political, not technical" (1987, 150).

Despite the wonder and excitement that surround these converging sciences, some influential people are calling for a halt to further development of certain types of technologies, and even limiting the pursuit of certain types of knowledge (Wishard 2001). These concerns are "based on the unknown potential of genetics, nanotechnology and robotics, driven by computers capable of infinite speeds, and the possible uncontrollable self-replication of these technologies this might pose" (230).

Technology Trends

Wishard (2001) identifies three trends that are moving us from one age to the next. First, in what we might recognize as globalization, he writes,

"For the first time in human history, the world is forging an awareness of our existence as a single entity" (226). The second trend is technology development, a phase that is "without precedent in the history of science and technology" (228). The third trend is "a long term spiritual and psychological reorientation that's increasingly generating uncertainty and instability" (230). Let's consider each of these in turn.

Globalization

Globalization is being seen in many ways. The economy has become a global economy. Education has gone global with studies abroad and online courses. American-based charities have taken up causes such as AIDS in Africa. Travel for business and pleasure now reaches virtually every part of this earth. These are just a few tangible examples of globalization.

Wishard (2001) made an interesting point about "global consciousness" when he said,

> If we're going to build a global age, it's got to be built on more than free markets and the Internet. Even more, it's got to be built on some view of life far broader than 'my nation,' 'my race,' or 'my religion' is the greatest. . . . It must have as its foundation some shared psychological and, ultimately, spiritual experience and expression. (228)

He believes terrorism is being fueled by this loss of identity, perhaps as groups grapple to hold on to their "norm."

The intellectual Paul Virilio is a skeptical critic of "techno-culture." His concern is that "techno-progress" is America's plan for "globalization"—something more like brain washing and totalitarian control (*The Information Bomb*, review). At the most basic level, skeptics and proponents agree that our definition of geography must now include a virtual landscape. "Within this cybernetic reconstruction of reality, the *global* becomes the center of things and the *local* the periphery, as virtual geography starts to dominate the real dimensions of the earth" (Virilio 2000, 10). "This domination is apparent in the construction of Internet communities, where the neighborhood unit is no longer local, but an elective, global association mediated by technology" (59).

Social Implications of Technology Development

Social theorists try to make sense of society, culture, and humanity in the world. Previously, location, race, religion, or even social status defined a group. Now sociologists have to redefine the groups that they study. With location a non-issue due to technology's impact on communications, and one's color, ethnicity, and social status veiled by a faceless persona, information technologies have changed the "groups" being studied. Mellor addresses this notion when he writes, "In some forms of sociology, 'culture' has come to replace 'society' as the central object of study . . . it has also allowed for the development of forms of reductionism where culture turns out to be determined by something deemed more fundamental, such as technology." He argues that this does not advance the field of sociology, and that inadequate theories "that ignore the human and religious dimensions of contemporary life in their intoxication with machine-mediated flows of information" should be replaced with, or at least include, "a proper engagement with human ontology" (357). Mellor points out that despite our dependence upon computers, televisions and mobile phones ". . . empirically-oriented studies have demonstrated the continuing sociological importance of embodied relationships with real people in specific, geographic locales" (359).

Technology's affect on individuals and society is indisputable. Messages are being sent that technology is good: it will improve medicine, boost the economy, enhance education, advance warfare, and will provide ample entertainment. On the other hand, messages are being sent that it is bad, even evil: it will erase sociology, eradicate religion, and replace humanity. The fear is that technology will harm the interdependence humans have on one another, and replace those relationships with machine-mediated relationships, or even worse, replace our dependence on a Higher Being with a reverence for the almighty machine.

> Technology in itself is neither good nor evil, but it does have the power to amplify the dispositions toward one or the other that exist in people, just as it amplifies the physical work they are able to accomplish. It does not have the power to make us other than what we are. Shaping who and what we are and determining how we direct our lives—which includes deciding the ends to which technology will be put—is the traditional role of morality, philosophy, and religion and the purpose of our social and political institutions. (Hogan 1997, 354)

As exciting as some of the breakthroughs in technology are, a deeper, more alarming question must be asked: Are the technologies we are creating improving the human condition or destroying humanity's meaning and significance altogether? (Wishard 2001). Even those technologies that "improve" the human condition by being faster, more accurate, and more convenient, are still adjusting what is our "norm." It is important to consider that technologies that do not alter us physically, medically, or cognitively, are still bound to affect us socially. This would logically necessitate new social and cultural norms.

There is an assumption that technology is progress, and progress is good, but are these changes actually for the better? Some theorists, philosophers, and technology critics feel that human beings are already being affected by the radical influx of information, technology, and change that has occurred. Wishard questions how we are to "live in a world that's changing faster than individuals and institutions can assimilate" (2001, 231). Conversely, Cardwell (1995) argues that one cannot enjoy the benefits of technologies such as modern medicine and public health without accepting the rest of science and technology.

And what of the wellness of our society? With all the advancements in the medical field, not to mention the "advancements" in all other domains of our existence, how is our society doing? Wishard gave some interesting statistics in his 2001 address. He said that the suicide rate among women has increased 200% in the last two decades; the teen suicide rate jumped 300% between 1960-1990; mental health is the fastest growing component of corporate health insurance programs; antidepressants are taken like aspirin; and rage is playing out far too often on the roads, in our schools, and in our families. Speaking on the acceleration of change and how much humans can take before we begin to show signs of distress, Wishard says, "The multiplying social pathologies indicate that individual and collective psychological integrity is already giving way" (2001, 229).

Ironically, some see a growing relationship between medicine and spirituality rather than a declining relationship. In an editorial debating Bainbridge's "Cyberimmortality: Science, Religion, and the Battle to Save Our Souls," Brenda Walsh points out that "more than half of the major medical schools in the United States have courses on spirituality and healing" (2006, 4). She also noted that a recent survey found that seventy percent of scientists reported themselves as "spiritual persons" and "believed that a basic truth guides our lives" (4).

The other side of the argument is that technology can have unimaginable consequences. Not only might the technologies developed harm us or our world, but more importantly, our way of life, our very social existence. Consider just one category of technological advances: the machinery of war. Many believe that war stimulates (and historically has stimulated) technology and invention. In fact, Kurzweil (1999) said that war is a true father of invention. Michael Marshall (2003) in an editorial writes,

> The advance of technology over the course of a century is astonishing. Sobering is the destructive power it has generated. As we contemplate the prospect of weapons of mass destruction in the hands of fanatic terrorists, or intercontinental ballistic missiles controlled by the eccentric dictatorial regime of North Korea, we are looking at a world where the many blessings of technology may be teetering on the brink of destruction.

The future of warfare could look like something from a science fiction movie. Neild and Pearson (2005) predict bacteria being used to detect explosives, bacteria used to break down explosives in mine fields, bacterial weapons, and wars fought by robotic soldiers.

To wonder what technological innovations might have been developed (or not developed) had there been no major wars in the twentieth century is a moot point. According to Cardwell (1995) it is unsound to assume that peace would mean fewer inventions. Cardwell theorizes that without war, the world would have seen more, albeit somewhat different, technological innovations. Many theorists argue that the technological innovations of war would have been discovered as a byproduct of a technological innovation for other purposes.

Mellor notes that several social theorists (Virilio, Castells, and Urry) have in common "the belief that a radical reconstruction of such things is taking place." What they find alarming is, "the dehumanization, disembodiment and moral anaesthetization that is now . . . accompanying the substitution of virtuality for reality" (2004, 364). Mellor cites Durkheim as being the social theorist whose theories are based on the notion that "being part of society is inextricably tied to our humanity, an idea of fundamental importance if we are to continue to study what societies really are, rather than succumbing to technologically driven fantasies about what they might be" (364).

Change has always produced fear, and perhaps that is why the most progressive scientists and theorists have been met with such opposition. However, if being part of a society is an inherent condition of being human, does that mean that our definition of society cannot change? If communication technology redefines time and place as variables in social interactions, does that really mean that we no longer need people? Look around you, and you see many examples of how technology is connecting people to the world. Technology has actually blessed *and* cursed us with the *inability* to be out of contact, in a world where email, instant messaging, telephones, cell phones, voice mail, caller id, beepers, pagers, and PDAs keep us constantly connected to the world around us. The definition of "connectedness" might be what is changing, rather than our need for it.

Spiritual Implications

> In the century and a half since Darwin, the truce between religion and science has been facilitated by the fragmentation of science into myriad little specializations. Each scientific subfield might present its own small challenges to faith, but to most people these might not add up to a serious intellectual challenge to religious belief. However, if the sciences become unified, religion may no longer be able to survive in the lacework of gaps that has existed until now . . . religion may feel a need to destroy science in order to save itself. (Bainbridge 2006, 28)

This unification is happening, as we see a convergence of the "big four"—nanotechnology, biotechnology, information technology, and cognitive science.

Beginning around 1980, researchers in sociology, psychology, and anthropology began developing rigorous models of religious faith. By attempting to explain it, they came perilously close to debunking it (Bainbridge 2006). Bainbridge warns of potential perils to religion as humanity makes intellectual advancements. "The authors of the Bible did not know that the Earth is a planet in orbit around the Sun, that the genetic code is carried by DNA molecules, or that the work of the brain is carried out by neurons. As our own ignorance diminishes, there is no guarantee that anything of the biblical worldview will survive" (2006, 29).

As science and technology advance, the very premise of our religious or spiritual doctrine—the idea that we are mortal and that we die and ascend to a heaven—is called into question. For example, take

Bainbridge's prospect for "cybernetic immortality"—the idea being that technology will evolve past just capturing images or video of our lives, but that it will capture our conscience, thoughts, ideas, memories, knowledge, and other personality characteristics held in our cognition that currently cannot be captured beyond our writings, scrapbooks, and images. This system would record, classify, and access all 50,000 episodic memories that a person potentially has. Bainbridge addresses the possibility that "converging technologies will offer humans extended lives within information systems, robots, or genetically engineered biological organisms" (2006, 25). This possibility involves rethinking human personality, from a complex, intangible set of characteristics and experiences to tangible, "dynamic patterns of information."

This also involves redefining human life. It would no longer end at the death of our physical bodies. But then again, those who believe in heaven believe life continues after the death of our physical body anyway. The difference involves the physical plain, here on earth, and what impact this might have on our existence here after our bodies have departed. Cyberimmortality is the possibility to extend our "life" here on earth. Bainbridge points out that "The convergence of cognitive science with information technology already threatens traditional beliefs that are the heart of religion, notable the need for God to save souls" (2006, 25). He writes that cyberimmortality "may meet resistance from religious groups arguing that the soul is a spirit, not a system" (25). Bainbridge contemplates that online memorials, which could include writings, images, or video, are the precursor to cyberimmortality.

Although much has been said about technology's potential impact on religion, less has been said about technology's impact on spirituality. Perhaps this is because spirituality cannot be quantified or measured. It is as unique as a fingerprint. Spirituality is individual; it exists anytime and anyplace. In that sense, spirituality has a likeness to the "virtual." Conversely, religion by its nature is a physical group, usually involves a set place (church, synagogue, mission, parish), and often a set time (Sunday school, worship service.) Therefore by its nature, religion (in its current state) does not lend itself to being "virtual." With that comparison, one can see how some theorists claim religion as a dated byproduct of the "information age."

"It is clearly the case that contemporary technological developments can have a significant impact upon social and cultural forms, and upon the ways in which people encounter and experience religious phenomena" (Mellor 2004, 358). Donald Cardwell, addressing the public perception of technology, has noted that technology is often blamed for the evils of the modern world. He writes that critics have blamed science (assumed to include technology) for the alleged lack of spirituality and declining religious beliefs, as well as for materialism (1995). Cardwell feels that these criticisms are unfounded, and points out that social critics, philosophers, novelists, and even politicians have done far more to attack religious beliefs than have technologists. Mellor notes a "general neglect of religious issues in accounts of the information age" (2004, 367) and that "the absence of a serious engagement with religion in much of the information society literature . . . is expressive of a lack of interest in such moral obligations. . . ." (369).

Wishard points out two Greek words that he feels might apply to our time: *hubris* (identifying ourselves with the gods, pride reaching beyond proper human limits) and, even stronger than hubris, *pleonexia* (an overwhelming resolve to reach beyond the limits, an insatiable greed for the unattainable.) He asserts that many of those in the forefront of the technological movement have crossed the line and have aligned themselves with gods, even *as* gods. Wishard points out stories throughout history (i.e., Adam and Eve) that all have the common theme, a warning that "limits exist on both human knowledge and endeavor; that to go beyond those limits is self-destructive" (2001, 229).

However, are theorists overreacting when they theorize the demise of religion and spirituality? To understand spirituality, one must understand the soul. This is a difficult thing for science to measure. Bainbridge writes,

> In *Descartes' Baby*, psychologist Paul Bloom argues that humans imagine that they have souls because the human brain has no awareness of its own functioning. We falsely perceive ourselves to be separate from our bodies. . . . Cognitive science is making considerable progress in understanding how human thought and behavior arise in the structures and electromechanical processes of the human brain, and it has found no evidence of a soul. (2006, 26)

Commenting on Bainbridge's article Andy van Roon writes, "The entire tone of (Bainbridge's) piece is that of a foregone conclusion suggesting that science will eventually erase the human need for, and therefore the very notions of, soul and the existence of a greater Being" (2006, 67).

That is analogous to putting a net in water, scooping, and when you realize that the net holds no water in it, theorizing that there is no such thing as water. We cannot define the soul by today's scientific methods. And perhaps cognitive science will answer questions about the spirit and the soul, and perhaps that is not where we need to look at all. Perhaps it is a personal definition that science cannot provide, or perhaps it is an all-new science, currently unknown to us, that will break the code of the human soul.

Interestingly, neuroscientists have discovered an area of the frontal lobe that appears to be activated during religious or spiritual experiences (Kurzweil 1999). Evolutionary biologists have postulated that there is a social utility for religious belief, so why would this not be reflected in our brain activity? This new discipline that aims to discover the "neural correlates of the divine" is often called "neurotheology" or "spiritual neuroscience" (Biello 2007). More recent research has found that no one particular "God spot" exists, but rather a neural network. This neurological, biological process does not belittle or scientifically negate the possibility of God or of the spiritual experience.

> No matter what neural correlates scientists may find, the results cannot prove or disprove the existence of God. Although atheists might argue that finding spirituality in the brain implies that religion is nothing more than divine delusion, the nuns were thrilled by their brain scans for precisely the opposite reason: they seemed to provide confirmation of God's interactions with them. (Biello)

Wishard speaks to this as well.

> [Q]uantum physics suggests there may exist some relationship between the human psyche and external matter. There may be some fundamental pattern of life common to both that is operating outside the understanding of contemporary science. In other words, we may be fooling around with phenomena that are, in fact, beyond human awareness; possibly even beyond the ability of humans to grasp. For at the heart of life is a great mystery which does not yield to rational interpretation. This eternal mystery induces a sense of wonder out of which all that humanity has of religion, art and science is born. The mystery is the giver of these gifts, and we only lose the gifts when we grasp at the mystery itself. (2001, 230)

Perhaps because of this quest for knowledge and understanding, there has been an increased interest in the unexplained, such as UFOs, ghosts and the paranormal. We seem to enjoy searching for meaning and understanding—the "mystery" alluded to by Wishard.

This seems to be an inherently human trait—searching for the solution to a puzzle that we secretly hope will never fully be understood. For once the code is broken there must be another puzzle to tease our knowledge-thirsty brains. Kurzweil notes that we are "more attached to the problems than to the solutions" (1999, 1). Hogan (1997) contends that the hostility that has greeted previous revolutionary thinkers (Copernicus, Darwin, Freud) comes not from any threat posed by the new way of thinking, but from a resistance to the demystification of our own humanity.

Something to consider—particularly for those who are concerned about the loss of religion and spirituality in this technologically advancing age—is what your Higher Being's plan is (God, Allah, or any other deity of your understanding). Do you believe there is a plan? Do you believe we have the power to alter that plan? Does creating intelligence usurp the power of God? Or are our actions part of the plan—creating our own evolution?

Many believe we are flying in the face of a Higher Being to develop the technologies we are developing. They not only believe we are headed down the wrong path, threatening our own souls, but that we are also threatening the future of humankind, even the world. This happened several times in biblical lore—for example, the Great Flood and Noah's Ark, where all sinners were washed away and the world "started over." So the argument still exists that we need to consider every step we make, and its potential current and future ramifications. All things must be considered, and resources need to be in place to resolve situations we *didn't* consider. We must act responsibly and prudently, for our own sake and the sake of our future.

The new quest might be to determine what it means to be *human*. Is it our thoughts, actions, relationships? Or is something immeasurable, like the soul? And how do we begin to understand something as unquantifiable as the soul, or spirituality in general? Technology operates on the notion that what is known can be improved upon. But how do we improve upon what we do not understand? Will we learn the significance of the soul only in reaction to the future we create using technology—when it is too late to go back?

Once we create artificial intelligence that can do anything humans can do (only faster), we may begin questioning the purpose and usefulness of humans. Yet no technology can replace or enhance human beings without a better understanding of what it means to be human. We must understand how we as humans think, emote, communicate, interact, and learn in order to improve upon that, let alone replace it with machines. As many technology theorists have noted (Wishard 2001), this also raises the question of why we need artificial intelligence with emotional and spiritual capability? The developing technologies, and the research into those technologies, might very well serve to teach us more about ourselves than we have ever before understood.

The Impact of Technology: Technology Issues

Technology has had an impact on most, if not all, of the mainstream topics of our time, including but not limited to economics, education, health and medicine, religion, communication, sociology, entertainment, marketing, military/warfare, agriculture, the environment, privacy and security, and politics. But there are other issues to consider regarding technology's impact: these are technology gaps and information overload. One of the foremost problems facing technology integration to the masses has been the gaps that form between gender, age, and socioeconomic status.

Traditionally, males excelled in math and the sciences more than their female peers (gender gap). According to the Department of Education website, the gender divide in computer use has been essentially eliminated as there is no overall difference between boys and girls in overall use of computers (U.S. Department of Education 2006). Cardwell notes that computer technology will likely be the first major technology in which women play a full part. "As a new technology it is without the established male hierarchies, assumptions and prejudices of older technologies" (507).

Youth more quickly adopt technologies as part of normal, everyday life, while more "seasoned" persons are slower to adopt. This is an age gap. Marc Prensky (2001) coined the term "digital natives" to describe those for whom digital technology has always been a part of life and "digital immigrants" to describe those who have had to adopt and adapt. He actually refers to this as the Immigrant/Native divide. We currently have digital immigrants teaching digital natives, using outdated methods to teach outdated content. This is not just an education issue, but a theoretical one regarding how the digital generation learns, processes, and adopts.

The socioeconomic gap is perhaps the most difficult to eliminate. Clear lines are perceived between the "haves" and "have nots," particularly in the school systems. The Department of Education has attempted to address this issue in its "No Child Left Behind" initiative. Many other benefits provided by technology (new medical procedures or medicines, for example) might only be accessible to a few.

One of the mixed blessings of the "information age" is the quick and relatively easy access to information that could not have been imagined just three decades ago. This has led some to conclude that "more is not better" since quantity does not always translate to quality. Information overload can happen as one searches information on the Internet (I got 1,830,000,000 results when I "Googled" *technology*. That approaches two *billion* links!)

Future Trends—2020

When we speak of technologies of the future, which are we discussing—the inevitable, or science fiction? There is a fine line between the two. What is currently impossible and material for science fiction, might very well become possible and be in development in 2020. For example, flying automobiles that have been imagined by dreamers and engineers for decades are actually now in development, utilizing GPS technology. Cars that parallel park themselves are already a reality. What we dream today may very well be tomorrow's reality.

Neild and Pearson (2005) have the enviable job of "Futurologist." They predict future technologies and the timeline in which they might occur. Although not scientific, it is fun to read their predictions, and time may prove them right as we move towards the year 2020. Examples of their predictions include: the first species brought back from extinction (2006-2010); patient records include multimedia (2008-2012); 3-D air display that has fingertip tracking (2008-2012); virtual reality overlays on real world (2008-2012); one's own tissue grows replacement organs (2011-2015); computer-enhanced dreaming (2011-2015); self diagnostic, self repairing robots (2016-2020); fully auto-piloted cars (2016-2020); holographic television (2020s); 3-D home printers (2020s); and robots that become physically and mentally superior to humans (2030s).

What are some potential benefits associated with the technologies of today and of the year 2020? Technology could serve to make the world a smaller place. The "us and them" that previously existed in the form of the cultural divide, could diminish as all cultures gain access to the same information. Things that previously could only be seen in person can be seen virtually. Things that formerly could not be seen with the naked eye, from the minute atom to the expanse of the cosmos, can now be seen. Medicines could be developed that cure currently incurable diseases. New delivery methods for medications could provide safer, less painful treatments. The economic potential of technology is mind-blowing.

As we move towards the year 2020, globalization will continue. Information will become ever more accessible. Privacy issues will become hotly debated. Computers will become indispensable, permeating every facet of daily life. Computer technology will continue to get smaller and faster, and we will see the merging of technologies as we have in our cell phones that act as calendars, calculators, and even email stations.

"Computers doubled in speed every three years at the beginning of the twentieth century, every two years in the 1950s and 1960s, and are now doubling in speed every twelve months. This trend will continue, with computers achieving the memory capacity and computing speed of the human brain by around the year 2020" (Kurzweil 1999, 3). In other words, by the year 2020, a computer will have a speed of about 20 million billion neural connection calculations per second. This is equal to the human brain. Supercomputers could reach this capacity by the year 2010.

Perhaps the biggest, most important steps technology will take by the year 2020 are in the fields of biotechnology, nanotechnology, and artificial intelligence. Biotechnology at its extreme would involve the eventual marriage of human and machine. This would mean that human beings will continue as a species, but not in a form we would recognize today. Scientific intellectuals call this the "Post-human Age." We would, in effect, be creating our own evolution.

Kurzweil calls human intelligence an evolutionary process, a billion years in the making.

> The brain's weakness is the extraordinary slow speed of its computing medium, a limitation that computers do not share with us. For this reason, DNA-based evolution will eventually have to be abandoned . . . indeed, evolution has found a way around the computational limi-

> tations of neural circuitry. Cleverly, it has created organisms that in turn invented a computational technology a million times faster than carbon-based neurons. (Kurzweil 1999, 101)

The emergence of a new form of intelligence that can compete with, and in fact exceed human intelligence, "will be a development of greater import than any of the events that have shaped human history. It will be no less important than the creation of the intelligence that created it" (1999, 5). Kurzweil writes that human intelligence, evolution's grandest creation, is "providing the means for the next stage of evolution, which is technology" (35). If this is the case, why are we in such a hurry to summon our evolutionary successors?

The potential applications of nanotechnology are fascinating, to say the least. The merger of nanomaterials with biology will result in new capsule medicines that can target specific diseases by being activated at the site of diseased cells by laser, ultrasound, or radio technology (Neild and Pearson 2005). Damaged tissues could be repaired or reproduced—replacing today's organ transplants—with the aid of tiny nano-material scaffolds that work to repair each individual cell. Nanotechnology could revolutionize environmental practices; waste water treatment, air purification and energy storage devices can be expected. Consumers may eat foods processed and packaged using nanotechnology anti-microbial agents. Household surfaces could be self-cleaning.

Artificial intelligence is another contender for the most theoretically-argued and debated technology of the future. What constitutes human intelligence? Can it actually be replicated and reproduced? And if so, for what purposes? Wishard says, "The experts tell us that by the year 2035, artificial robotic intelligence will surpass human intelligence. . . . And a decade after that, we shall have a robot with all the emotional and spiritual sensitivities of a human being" (2001, 228). For what purpose would we want a human-machine robot that can act exactly as we act? Bainbridge notes that even as computer scientists create "technologies that effectively imitate more and more distinct functions of human intelligence . . . the goal of full artificial intelligence remains elusive" (2006, 26).

"Once computers are as complex as the human brain, and can match the human brain in subtlety and complexity of thought, are we to consider them conscious? . . . Some philosophers believe it is not a meaningful question, others believe it is the only meaningful question in philosophy" (Kurzweil 1999, 6). This debate is bound to get more complex as machines appear to have emotions, free will, and even spiritual experiences. Kurzweil argues that people will believe them.

Also debated is the potential of humans to download the contents of their brains to a computer. Assuming a computer existed with sufficient capacity, it theoretically could be done by scanning one's brain, then recreating the connections and circuitry on a neural computer. Kurzweil ponders how well this would work. As with any new technology, he points out the imperfections that might exist, resulting in imperfect and imprecise downloads.

Interesting are those questions that begin to arise regarding the ramifications of the very future we predict. If a human-machine were to be created, then what happens to reproduction if humanoids live in perpetuity? Will the Earth's land surface area reach standing room only with so many immortals about? (Albright 2006). What if AI avatars create their own civilization?

If humans and machines were blended into a seamless AI being, would humans not be able to detect their technologically advanced counterparts? This could bring discrimination to a whole new level as humans try to compete with beings designed to be superior. Carrying this to an extreme, one could predict a backlash. Neild and Pearson (2005) make a prediction for the year 2015 that the replacement of people leads to an antitechnology subculture. This thought is seconded by Kurzweil who points out that although the Luddite movement is currently unfashionable, it lingers not far below the surface and "will come back with a vengeance in the twenty-first century" (1999, 86). Much the same as witches were sought out in the 1600s in Salem, Massachusetts, AI beings might be sought out in our future. It is actually a scary prospect, that we might create beings that we then reject. And if we reject them, would they be sufficiently superior to eliminate us first? Bainbridge writes,

> It is remarkable that substantial opposition to robots, computers, and information systems has not already arisen. There has yet to be any outcry comparable to the religiously based movement to ban human reproductive cloning. Computers have evolved so gradually, and serve so many millions of people so well, that few people feel threatened. But that may quickly change when new reports begin describing robots or AI avatars said to carry the personalities of deceased persons. (2006, 29)

Until the scientific community and the religious community can agree upon when life begins, we will continue to have religious groups arguing against medical technologies and procedures like stem cell research. Until there is an understanding of what it means to be human, technologies that encroach upon and accurately approximate humanity will be feared. What are the things that make us most human? Is it our physicality? Our DNA? Is it our emotions? Is it our knowledge and quest for knowledge? Is it our social cohesiveness? Or is it our need to feel loved, useful, and valued? What happens if we lose sight of any one of these aspects of ourselves?

As we create machines that can do what we do—but faster and better—are we eliminating ourselves? We need an understanding of what it means to be human, what we like about ourselves as people, those things that we would not want to have replaced. We need an understanding of who we are, what we are, what we want to keep, what we want to shed, in order to save ourselves. We need an understanding of ourselves before we make predictions regarding how current and future technologies might affect us.

Do we not ascribe to ourselves the level of a God when we believe we have the power to change humanity, the world, and life as we know it? History has proven that humans can do a great deal of harm—to each other, to themselves, and to the earth. And what about the potential humans have for doing good—for *creating* good? To believe we have the power to change God's will is a heretical presumption. It is dangerous to believe we are not expected to make informed, responsible decisions regarding ALL things in life, including technology. But what if God's plan includes the development of technology?

Predicting the future of technology with any degree of accuracy is practically impossible due to the complexity of interacting influences (Cardwell 1995). What is clear is that the future of technology demands that we gain an understanding of what it means to be human. The future of humanity demands that we gain an understanding of where we as humans fit into life, our planet, and even our universe. Many theorists, theologians, authors, scientists, and even religious figures, question the future of spirituality in an age where technology may reproduce the very relationships, environments and symbols that we believe are tied to our faith. They question the future of spirituality in an age where technology disproves long-held beliefs about cognition, the grandness of the cosmos, the meaning of "life", and the finality of death.

There may come a day when we gain an understanding into ourselves and previous, deeply-held beliefs are proven wrong. Does this mean the end of religion? Does this signal the end of spirituality? To answer this, we must determine what spirituality, even religion, offers us. My belief is that religion offers us hope. Hope that there is more to life than the minute amount of time we are on this earth. Hope that we ascend to a better place (i.e., Heaven) and are reunited with those we love. Regardless of what religious beliefs one holds, and how technology may be used to quantifiably prove or disprove those religious beliefs, humanity's need for hope cannot be replaced, no matter what the state of technology.

Perhaps the biggest danger, as we move towards the year 2020, is that humans will put their hope in technology itself and abandon traditional religion and higher principles. Perhaps an argument could be made that religion will not be abandoned, but instead be reinvented, from something that competes with science for the souls of mortals to something that embraces science and its effects on our souls. Will God be redefined? Will Heaven be redefined? Will our souls be redefined? Will the meaning of being human be redefined? Only time will tell whether these hopes and fears are valid, or just science fiction.

References

Albright, Thomas B. 2006. Answers Needed on Human-Machine Intelligence [Letter to the Editor]. *The Futurist*, May–June.

"Artificial intelligence." 2006. *Britannica Concise Encyclopedia. Answers.com.* http://www.answers.com/topic/artificial-intelligence. (accessed October 17, 2006).

Austin, Chad. July 14, 2004. Study Shows Americans Encouraged by Prospects of Nanotechnology. News Release: North Carolina State University. http://www.ncsu.edu/news/press_releases/04_07/211.htm. (accessed on October 27, 2006).

Bainbridge, William Sims. Cyberimmortality: Science, Religion, and the Battle to Save Our Souls. 2006. *The Futurist*, March–April.

Bainbridge, William Sims. 2006. [Letter to the Editor.] *The Futurist*, July–August.

Baker, Monya. 2006. "Will Work for Food." *Wired*, December.

Biello, David. 2007. Searching for God in the Brain. *Scientific American*, October. http://www.sciam.com/article.cfm?id=searching-for-god-in-the-brain&print=true. (accessed October 1, 2008).

"Biotechnology." 2003. *The Columbia Electronic Encyclopedia,* Sixth Edition. http://www.answers.com/topic/biotechnology. (accessed October 17, 2006).

Cardwell, Donald. 1995. *The Norton History of Technology*. New York: W. W. Norton.

Convention on Biological Diversity. 2005. Cartagena Protocol on Biosafety: Frequently Asked Questions on the Biosafety Protocol. https://www.cbd.int/biosafety/faqs.asp?area=biotechnology&faq=5. (accessed October 27, 2006).

Hirsch, E. D., Jr. *Cultural Literacy*. 1987. Boston: Houghton Mifflin.

Hogan, James P. 1997. *Mind Matters: Exploring the World of Artificial Intelligence*. New York: The Ballantine Publishing Group.

Kurzweil, Ray. 1999. *The Age of Spiritual Machines*. New York: Viking Penguin.

Marshall, Michael. 2003. Destructive power of technology (Editorial). *World and I* 18.4 April: 7. *Expanded Academic ASAP*. Thomson Gale. Virginia Tech. http://find.galegroup.com.ezproxy.lib.vt.edu:8080/ips/infomark.do?&contentSet=IAC-Documents&type=retrieve&tabID=T003&prodId=IPS&docId =A100839389&source=gale&srcprod=EAIM&userGroupName=viva_vpi&version=1.0. (accessed on October 4, 2006).

Mellor, Phillip A. 2004. Religion, Culture and Society in the 'Information Age'. *Sociology of Religion* 65:4.

"Nanotechnology." 2006. *Computer Desktop Encyclopedia*. Computer Language Company. http://www.answers.com/topic/nanotechnology. (accessed October 17, 2006).

Neild, Ian and Ian Pearson, eds. 2005. *BT Technology Timeline*. August. http://www.btinternet.com/~ian.pearson/web/future/2005 timeline.doc. (accessed October 26, 2006).

Neild, Ian and Ian Pearson. 2006. A Timeline for Technology: To the Year 2030 and Beyond. *The Futurist*, March–April.

Prensky, Marc. 2001. Digital Natives, Digital Immigrants. *On the Horizon,* October. http://www.marcprensky.com/writing/Prensky—Digital Natives, Digital Immigrants—Part1.pdf. (accessed September 30, 2008).

"Technology." Merriam-Webster Online Dictionary (based on the print version of *Merriam-Webster's Collegiate® Dictionary, Eleventh Edition.*) http://www.m-w.com/dictionary/technology. (accessed on October 3, 2006).

The Information Bomb. 2000 (Review)(Brief Article). *Publishers Weekly*, June 19. http://ezproxy.lib.vt.edu:8080/login?url=http://search.ebscohost.com/login.aspx?direct=true&db=ehh&AN=3275552&site=ehost-live&scope=site. (accessed October 1, 2008).

U.S. Department of Education. Educational Technology Fact Sheet. March 29, 2006. http://www.ed.gov/about/offices/list/os/technology/facts.html. (accessed October 27, 2006).

U.S. Patent and Trademark Office. March 2006. All Technologies Report. http://www.uspto.gov/go/taf/all_tech.pdf. (accessed October 24, 2006).

Van Roon, Andy. 2006. The Limitations of Science in Measuring a Soul [Letter to the Editor]. *The Futurist*, July–August.

Vinge, Vernor. 1993. The Coming Technological Singularity: How to Survive in the Post-Human Era. http://www-rohan.sdsu.edu/faculty/vinge/misc/WER2.html. (accessed February 11, 2007).

Virilio, Paul. 2000. *The Information Bomb*. London: Verso.

Walsh, Brenda. 2006. Finding a Role for Religion in the Scientific Age [Letter to the Editor]. *The Futurist*, May–June.

Wishard, William Van Dusen. 2001. Between Two Ages. Address to Coudert Institute. Reprinted in *Vital Speeches of the Day,* January 15, 2002. Mt. Pleasant, SC: City News Publishing Company.

Index

About the Editor

Robert L. Menz, DMin, is the Employee Counselor and Corporate Employee Assistance Director for Emerson Climate Technologies in Sidney, Ohio, director of the Shelby County Chaplaincy Board in Sidney, and an adjunct faculty member at Edison Community College in Piqua, Ohio. Dr. Menz is a Fellow in the American Association of Pastoral Counselors, a Diplomate in the American Psychotherapy Association, and a Diplomate in the College of Pastoral Supervision and Psychotherapy. He is board certified by the Association of Professional Chaplains, and a Certified Employee Assistance Professional in the Employee Assistance Professionals Association. He is author of *A Pastoral Counselor's Model for Wellness in the Workplace: Psychergonomics,* and *A Memoir of a Pastoral Counseling Practice*—both from Haworth Press—as well as numerous articles in professional journals. Robert and his wife Ruth together have four children and thoroughly enjoy their six grandchildren.

About the Authors

Clinical psychologist **James E. Gebhart**, PhD, in addition to having been in private practice in Columbus, Ohio for twenty-five years, is an ordained Elder of the United Methodist Church, and fully endorsed by the Division of Chaplains and Related Ministries of the Board of Higher Education and Ministry. Dr. Gebhart has extensive education in philosophy, theology, education, and psychology, having received his PhD from The Ohio State University in 1982. He is a Diplomate of the College of Pastoral Psychotherapy and Supervision, and of the American Association of Pastoral Counselors. A member of the American Academy of Psychotherapists, he is Certified as a Supervisor by the Association for

Clinical Pastoral Education. For a number of years he held special appointment to the faculty of The Methodist Theological School in Ohio. Dr. Gebhart is currently the Immediate Past President of the College of Pastoral Psychotherapy and Supervision and was the tenth president of the Association for Clinical Pastoral Education. He founded the Ohio Institute of Pastoral Care, was co-founder of Gebhart and Associates (Consultants to Business in Corporate Change), and a founding representative to the Congress on Ministry in Specialized Settings. He has served the American Red Cross as a member of the Spiritual Aviation Incident Response Team. His articles have appeared in the *Journal of Pastoral Care and Counseling*.

Roy E. Godwin, PhD, has served as an ordained minister in both the Southern Baptist Convention and the American Baptist Churches, USA, for over forty years, mostly in metropolitan areas of the nation. His longest ministry—nearly twenty years—was service as a denominational executive in the metropolitan Washington, DC area. Dr. Godwin has had a varied ministry as pastor, professor, denominational administrator, missionary, chaplain, consultant, and corporate and personal coach. He has been actively engaged in ecumenical and interfaith enterprises that have brought significant benefits to persons, organizations, and denominational endeavors.

Suellen Mazurowski, JD, a retired attorney, lives in Bluffton, South Carolina. Her experience includes several years of teaching in the public schools of Pennsylvania and Ohio. After obtaining her Juris Doctor degree, she served as Assistant Attorney General for the State of Ohio, Assistant Prosecuting Attorney for Clark and Warren counties, and Director of the Clark County Child Support Enforcement Agency. While working in that county, Ms. Mazurowski represented many school districts. From 1995-2004, Suellen engaged in private law practice in her own firm in Sidney, Ohio. Since retirement she has been employed as a legal consultant. Ms. Mazurowski continues to be a member of the Ohio Bar, belongs to the Ohio State Bar Association, and is currently studying for the South Carolina Bar Examination. Suellen has three adult children and two stepdaughters. Between them, Suellen and her husband Norm Mazurowski have ten grandchildren.

Janice Michael is dean of the Darke County Campus of Edison Community College, Greenville, Ohio, where she served as an adjunct faculty member in business and economics since 1983. She holds a bachelor of arts degree in history from Ohio Northern University and a master's of science in administration from Central Michigan University.

Gwendolyn Ogle received her PhD from the College of Human Resources and Education at Virginia Tech in 2002. Her dissertation focused on designing software to streamline and manage the formative evaluation process. Trained in education, instructional design, and educational technology, she emerged from the Instructional Technology program with a passion for technology and a skill set in the evaluation of educational technologies. A three-year fellowship with NASA allowed her to specialize in the evaluation of technologically-enhanced educational resources. She worked as an independent consultant in educational evaluation and instructional design for five years during and after completing her PhD. In 2005, Dr. Ogle incorporated ID & E Solutions, a small consulting firm specializing in educational evaluation and instructional design. Dr. Ogle is married to Dr. Todd Ogle. They have three children, Jeffrey, Hailey, and Gracelyn.

Robert Strayer is Professor of Business and Economics at Edison Community College, Piqua, Ohio, where he has taught for thirty-one years. He does consulting work for a financial services company and is president of a farm corporation. He has earned two Bachelor of Science degrees, and a Master's degree in business from The Ohio State University. In addition, he holds a Master of Science degree in financial services from The American College at Bryn Mawr, Pennsylvania. Mr. Strayer is a Chartered Life Underwriter and Certified Financial Planner.

Brian C. Thomas, JD, an attorney with Graydon, Head & Ritchey LLP in Cincinnati, Ohio, specializes in employment and workers' compensation issues. He is a member of the Cincinnati Bar Association, the board of trustees of People Working Cooperatively, and serves as the Vice-President of the Black Lawyers Association of Cincinnati. He is author of Examining a Beggar's First Amendment Right to Beg in an Era of Anti-Begging Ordinances: The Presence and Persistence Test, published in the *University of Dayton Law Review*. He received his Bachelor of Science degree in management, with honors, from the Georgia Institute of Technology and his Juris Doctorate, *cum laude*, from the University of Dayton School of Law, where he was a member of the moot court team and Editor in Chief of the law review.